INSTITUTE OF GEOLOGICAL SCIENCES
Natural Environment Research Council

OVERSEAS MEMOIR 3

The geology of the Maya Mountains, Belize

J. H. Bateson and I. H. S. Hall

London Her Majesty's Stationery Office 1977

ISBN 0 11 880765 X *

CONTENTS

TABLES

MAP

ABSTRACT

The Maya Mountains of Belize represent an uplifted fault block of pre-Cretaceous basement which is geographically isolated from the larger areas of Palaeozoic rocks in Guatemala, Honduras and Mexico. A number of travellers and scientists have made observations on the geology of these mountains since the end of the nineteenth century. The first geological account accompanied by a map was compiled by L. H. Ower (1928). The next major work was done between 1951 and 1955 by C. G. Dixon (1956) who produced a revised geological and economic account of the area. Basically he defined two sedimentary groups separated by an interval during which granites were intruded and widespread erosion took place. Dixon was able to give a fairly accurate age to the younger of his two sedimentary rock groups, which he named the Macal Series. He considered that it extended from the Pennsylvanian to the Middle Permian. The older group, which he named the Maya Series, yielded no fossils and was believed by Dixon to be much more metamorphosed and deformed. This he placed in the Lower Palaeozoic.

Recent investigations have not upheld Dixon's subdivision of the sediments into two separate series. They are now all assigned to one group which has been correlated with the Santa Rosa Group of Guatemala and has been given the same name. The Santa Rosa Group in the Maya Mountains of Belize has been intruded by granites and contains a contemporaneous facies of lavas and associated sediments which has been called the Bladen Volcanic Member.

Structural, palaeontological and radiometric data are presented to support these concepts, and an attempt has been made to redefine the palaeogeography in the context of current ideas on the Palaeozoic of adjoining areas of north Central America.

A synopsis is presented of the available data on the potential economic mineral deposits of the area. In many instances insufficient work has been done on them for a reliable estimate of their potential to be made. Data obtained from the 1969–70 geochemical survey does, however, seem to indicate that the area as a whole presents little hope of major economic mineral deposits being discovered.

Bibliographical reference

Bateson, J. H., and Hall, I. H. S. 1976.
The geology of the Maya Mountains, Belize.
Overseas Mem. Inst. Geol. Sci. No. 3.

Authors

J. H. Bateson, BSc, MIMM
Institute of Geological Sciences, London

I. H. S. Hall, BSc
Institute of Geological Sciences, Edinburgh

ONE
Introduction

The territory of Belize, formerly British Honduras, faces the Caribbean and occupies a strip of the Central American coastline approximately 290 km from north to south, and 100 km, maximum, from east to west. It adjoins the Republic of Mexico on the north and Guatemala on the west and south.

PHYSICAL FEATURES

The Maya Mountains (Plate 1) form the only extensive area of high ground in Belize and lie between longitudes 88° 30′ and 89° W and latitudes 16° 30′ and 17° 05′ N. Peaks and ridges commonly rise to 900 m (3000 ft) with Victoria Peak (16° 48′ N, 88° 37′ W) in the Cockscomb Range (16° 49′ N, 88° 37′ W) reaching an elevation of approximately 1095 m (3650 ft)—the highest point in Belize. The mountain mass is deeply dissected by many rivers whose courses are controlled in many places by jointing or faulting. The main watershed forms an arc concave to the west, which extends eastwards from Baldy Beacon (17° 00′ N, 88° 47′ W) to Baldy Sibun (17° 00′ N, 88° 45′ W) and thence southwards to the Cockscomb Range (16° 49′ N, 88° 37′ W) before curving to the south-west and passing to the north of the Bladen Branch (16° 31′ N, 88° 51′ W) and westwards into Guatemala.

Although the Maya Mountains form a separate identifiable topographic unit with respect to the rest of the territory they are themselves composed of a number of different topographic expressions directly related to the underlying geology.

The areas underlain by granite have a relatively subdued topography consisting of broad rounded hills. The numerous streams occupy shallow open valleys many of which are closely controlled by the jointing pattern. Most of the granite is covered with thick tropical forest, but in the area of Mountain Pine Ridge the vegetation on the granite is a sparse mixed forest of pine and oak with large areas of open grassland. The origin of this vegetation type is open to much speculation but it has been suggested that recurrent fires over a long period of historical time have reduced the natural soil cover by exposure to active weathering and erosion so that the main species to survive are those relatively resistant to fire and able to thrive on poor soil types.

HISTORY OF INVESTIGATIONS

The earliest geological observations in Belize are recorded in the diary and subsequent report of Henry Fowler (1879). In 1878 he made the first crossing of the Maya Mountains in modern times during which he kept a note of the rocks he encountered. A few years later, in 1885, G. H. Wilson traversed several of the larger rivers but unfortunately he died before he could produce a complete document on his travels. His maps and notes are, however, available to later workers in this field. Of these Sapper (1889) was the first to undertake detailed investigations. He recognised, in addition to granites, the metamorphic and fossiliferous Palaeozoic sediments which he regarded as facies of one main group. In this belief he was followed by Ower who undertook field work between 1921 and 1926, and produced the first geological map of British Honduras in 1928. He regarded the schists and other metasediments as metamorphosed equivalents of the fossiliferous Palaeozoics and expressed the view that the rocks classed as 'porphyry' included flows and probably ash beds, which are, by inference, of approximately the same age as the associated sediments (Ower, 1928). Like Sapper before him, Ower attempted some correlation, particularly of the post-Cretaceous rocks, with those of neighbouring Guatemala, but many of his conclusions in this field were modified in later years by G. Flores (1952), a petroleum geologist whose particular interest lay in the younger, potentially petroliferous beds.

All these sources of information were used by C. G. Dixon, who from 1951 to 1955 was seconded from the Geological

Plate 1 Panorama of the southern Maya Mountains
The flat-lying savannah-covered Recent deposits in the foreground contrast strongly with the densely forested hills of Cretaceous limestone and the more distant mountains of Santa Rosa Group sediments and volcanics.

Survey of British Guiana (now the Republic of Guyana) to prepare a new geological map and report of southern British Honduras, paying particular attention to the investigation of the mineral potential of the area. The presence of alluvial tin and gold had been indicated by the work of Sapper and confirmed by Ower who also identified molybdenite and monazite in some of the pan concentrates. Dixon added considerably to the knowledge of the occurrence of each of these minerals by pitting and detailed prospection. He also located and examined in some detail a barytes vein in the Mountain Pine Ridge (16° 59′ N, 88° 56′ W) to the south of Augustine (16° 58′ N, 89° 00′ W) and completed some preliminary work on the limestones and dolomites of the Cretaceous.

Since Dixon's publication (1956) no further geological work was carried out in the area until 1967, when the Canadian consultants, D. R. Derry Ltd, negotiated an exclusive prospecting permission which covered most of the Maya Mountains area. In addition to collecting samples for geochemical analyses for copper and zinc the geologists of this company made some observations on the general geology which appear in the Company's annual reports to the Government of British Honduras. They concluded their work in the area in 1969.

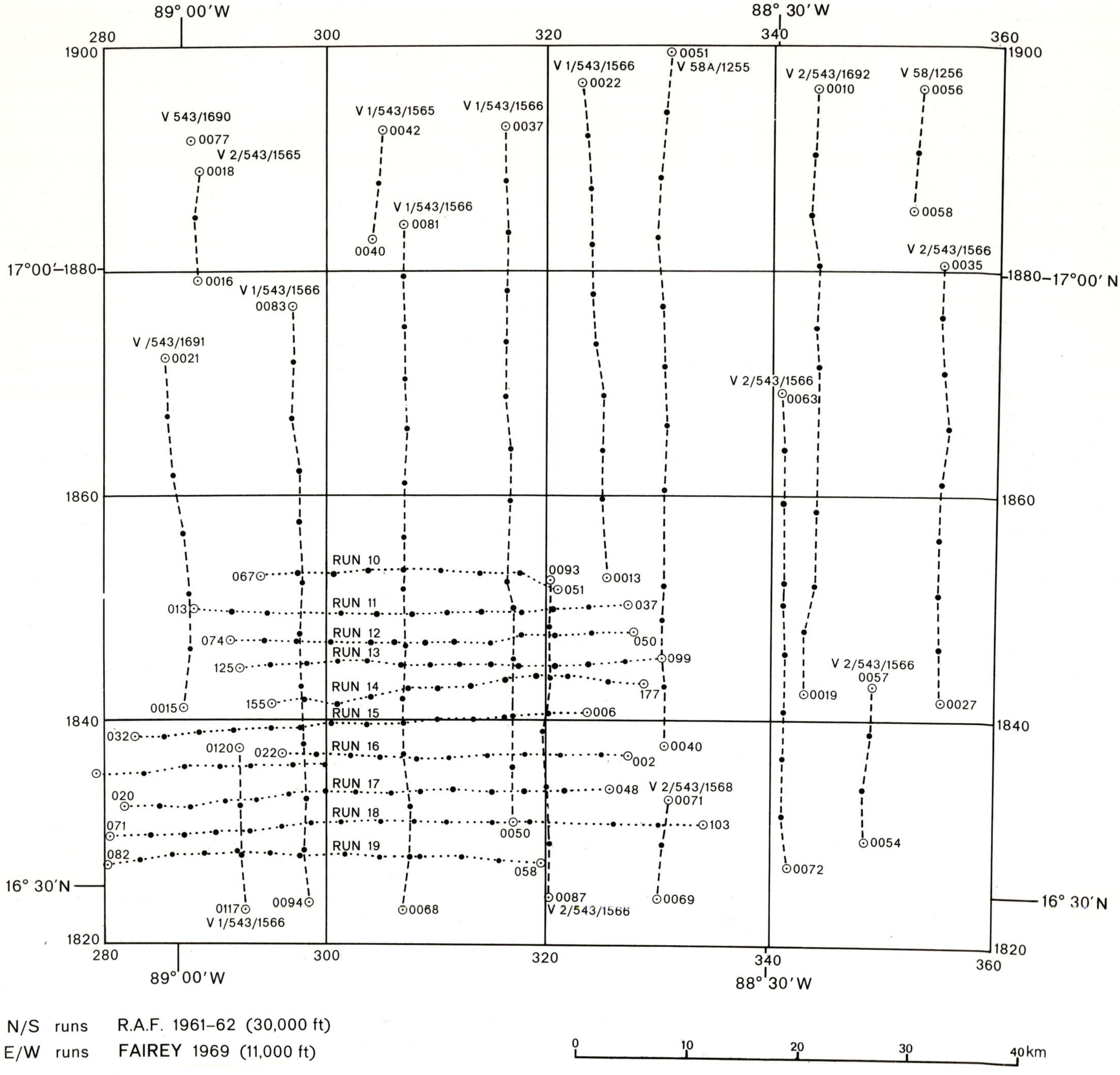

Figure 1 Flight lines for aerial photography

The Institute of Geological Sciences was asked by the Government of British Honduras to revise the existing geological map and to undertake a further programme of geochemical reconnaissance in the Maya Mountains. Consequently the Ministry of Overseas Development and the Government of British Honduras jointly bore the costs of field and laboratory work which was undertaken in 1969 and 1970. The results of this survey form the basis of this memoir. The authors have drawn fully and freely upon all the earlier work and have made considerable use of the aerial photography of the region, particularly that flown in 1969 (Figure 1).

SUMMARY OF GEOLOGY

The Maya Mountains occupy a position of considerable interest in the geology of this portion of Central America since they represent the only extensive outcrop of Palaeozoic rocks to the north of those of central Guatemala. Correlation, however, is made difficult because of the much higher degree of metamorphism exhibited in the Guatemalan rocks. By far the greater part of the Maya Mountains is composed of variably metamorphosed sediments which have been intruded by a number of mineralogically similar granitic stocks. Lavas and tuffs, which the available evidence suggests are contemporaneous with the main sedimentation of the area, occur in the southern portion of this mountain block.

In this paper the view is expressed that the sediments and volcanic rocks of the Maya Mountains comprise one geological group extending in time from the Pennsylvanian to approximately Middle Permian, and were intruded by granitic rocks during early Triassic times. New palaeontological data afford closer comparison between some of the rocks of the Maya Mountains and similar sediments in the neighbouring territories. They also support the concept of a single geological group as distinct from the two-fold division proposed by Dixon (1956). No geological record of the time between the Triassic and the Cretaceous has been preserved, and it is likely that during much of this interval the area was undergoing extensive denudation.

TWO

Palaeozoic sediments and extrusive rocks

THE SANTA ROSA GROUP

Dixon proposed a two-fold division of the sequence of sediments and low-grade metasediments that form the largest part of the Maya Mountains (Dixon, 1956). His older sediments, the Maya Series, were, he claimed, the more metamorphosed of the sediments and exhibited tight isoclinal folding. These were unconformably overlain by a series of unaltered open-folded sediments, containing characteristic thin limestones, which he named the Macal Series after the type section exposed in the Macal River valley (16° 45′ N, 88° 51′ W). According to Dixon the volcanics of the Bladen area were an intrusive phase later than the sedimentation and, since he had established that the youngest of the sediments was Middle Permian, he assigned a post-Middle Permian age to the so-called Bladen Porphyry.

The present authors have published elsewhere a revised nomenclature for the rocks of the Maya Mountains together with a much revised time scale based upon new palaeontological data and radiometric dating of the granites and volcanics (Bateson and Hall, 1971). The new evidence does not substantiate the previous subdivisions of Dixon and consequently the following new names have been designated. The sediments comprising Dixon's Maya Series and Macal Series are now combined into the Santa Rosa Group, and the Bladen Porphyry—renamed the Bladen Volcanic Member—is also considered to be a part of the Group (Figure 2). The evidence from new palaeontological determinations and from the radiometric dating of the intrusive granites and volcanics, together with detailed structural analysis, supports the conclusion that there was continuous sedimentation from at least the Pennsylvanian to the Middle Permian. The localised volcanics of the Bladen Volcanic Member form an integral part of the Santa Rosa Group and were extruded probably about the end of the Pennsylvanian or early part of the Permian.

The rocks forming this Group occupy about 80 per cent of the total area of the Maya Mountains, the remaining 20 per cent being underlain by the later granites. In the extreme west both granites and sediments are overstepped by Cretaceous limestones. Generally the surface expression of sedimentary rocks of the Santa Rosa Group is broken country with deeply incised streams, but the topography may be modified locally by a particular lithological type. Sandstones, for example, tend to produce the higher rolling hills with steep scarp features, while the finer-grained rock types, for example mudstones, slates, and quartzites, tend to produce flat-topped ridge and valley type topography at no great elevation with a well-developed 'herring bone' drainage pattern.

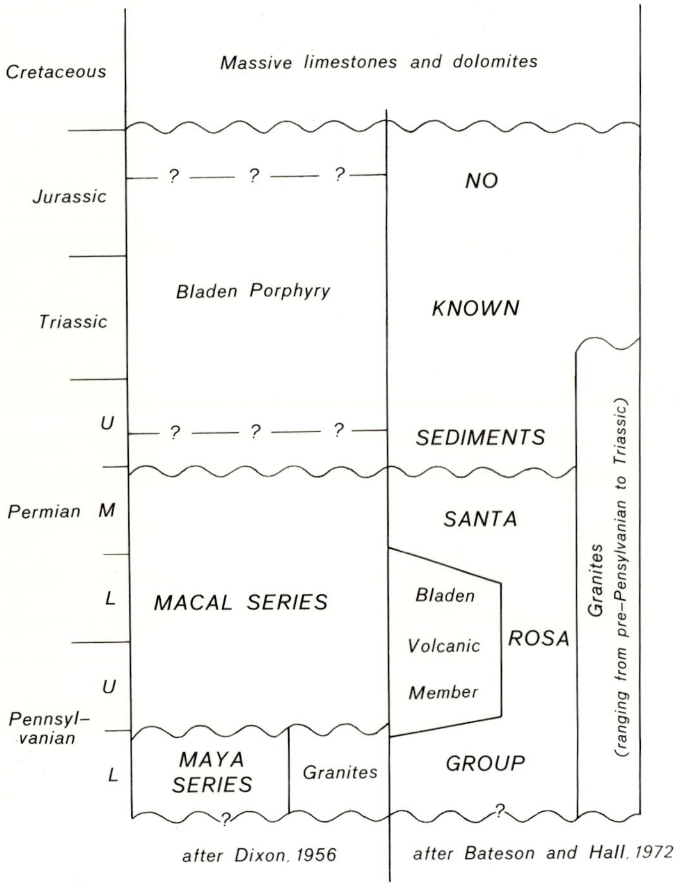

Figure 2 Revised pre-Cretaceous stratigraphy, Maya Mountains, Belize

Argillaceous and arenaceous rocks occur in the Santa Rosa Group with noteworthy crinoidal limestones and calcareous mudstones. The generalised distribution of the outcrop of these major lithological types is indicated in Figure 3. The dominantly arenaceous beds occupy three major areas: in the north, the Baldy Beacon–Baldy Sibun area (17° 00′ N, 88° 46′ W) and its possible extension on the southern flank of the Mountain Pine granite (16° 59′ N, 88° 56′ W); a more central area close to the confluence of the Raspaculo River and Monkey Tail Branch (16° 47′ N, 88° 55′ W); and in the south, flanking the sediments and volcanics of the Bladen Volcanic Member. (The significance of these areas is discussed later.) There is some evidence also that the sediments in the headwaters of the South Branch (17° 00′ N, 88° 32′ W) of the North Stann Creek and the Blackwater Branch (16° 55′ N, 88° 32′ W) contain an important arenaceous sequence. The rest of the Maya Mountains is composed of dominantly fine-grained sediments with some limestone which are believed to include the youngest of the Palaeozoic rocks of this area (that is, Middle Permian).

Arenaceous and rudaceous rocks

The most spectacular of these rocks are coarse conglomerates that are well exposed along the Baldy Beacon–Baldy Sibun road (Plate 3.2). The pebbles in these conglomerates are composed of hard sandstone and quartzite, with some phyllite and contact-metamorphosed sediments. They vary greatly in size and are up to 30 cm in diameter. The conglomerates are characteristically poorly sorted. Locally the conglomerate units may be 5 m thick; more usually they are about 1 m thick and separated by lenses and bands of finer sandy and silty material. The pebbles are well rounded and in places show an imbricate arrangement. The matrix of the conglomerate is normally composed of sub-rounded quartz and hard sedimentary fragments up to 0.5 mm in size, and is often iron-stained. Polymict conglomerates are an important part of the succession in the Baldy Beacon–Baldy Sibun area (16° 59′ N, 88° 46′ W) and in the area of the Trio Branch (16° 34′ N, 88° 39′ W). They also occur in the headwaters of the South Branch (16° 59′ N, 88° 33′ W) where they are closely associated with fossiliferous mudstones. Transported boulders of polymict conglomerate are found in the beds of many streams flowing into the North Stann Creek (16° 59′ N, 88° 20′ W) from the high ground to the south.

The outcrop of the coarse conglomerates at Baldy Beacon provided Dixon (1956) with field evidence on which he based his conclusion that there was an unconformable relationship between the Maya Series and the overlying Macal Series. There is, in some places, an angular nonsequence between the base of a conglomeratic band and the top of the underlying sediments, but this relationship is interpreted here as being consistent with the manner of deposition of these coarse deposits. No evidence was seen in the field which indicated the existence of a major regional unconformity between the conglomerate units and the earlier sediments. In the area to the north of Baldy Beacon (17° 00′ N, 88° 49′ W), for instance, bands and lenses of conglomerate do not show any unconformable relationship with the beds with which they are in contact.

The conglomerates are associated with sandstones (Plate 2.1 and 2.2), most of which, and particularly those of a medium to coarse grain size (average grain size greater than 0.3 mm), show a poor degree of sorting with the groundmass consisting of fine granular quartz usually with much sericite and occasionally muscovite. The coarser grains in these rocks are up to 1 mm in diameter but generally show only a minor degree of rounding. Quartz grains are dominant, but in most of the rocks there are also very variable amounts of feldspar and rock fragments.

These rocks represent sediments which accumulated rapidly with a minimum time for sorting and for chemical disintegration. Most of the coarser sandstones are subgreywackes, but sorting improves with reduction in grain size and some of the fine, well-sorted sandstones would properly be classified as orthoquartzites. These rocks are characteristic of the subgreywackes (Pettijohn, 1957) in that they do not show very many sedimentary structural features. Some small-scale current bedding has been observed as has the local development of ripple markings. Grading of beds was also recorded in those areas where there are sufficiently good exposures to enable the transition from sandstone to mudstone to be observed. Similarly the coarser and usually less well-sorted of these sediments may lie on the channelled surface of the finer beds beneath, and in these conditions

Figure 3 Lithofacies map of the Santa Rosa Group

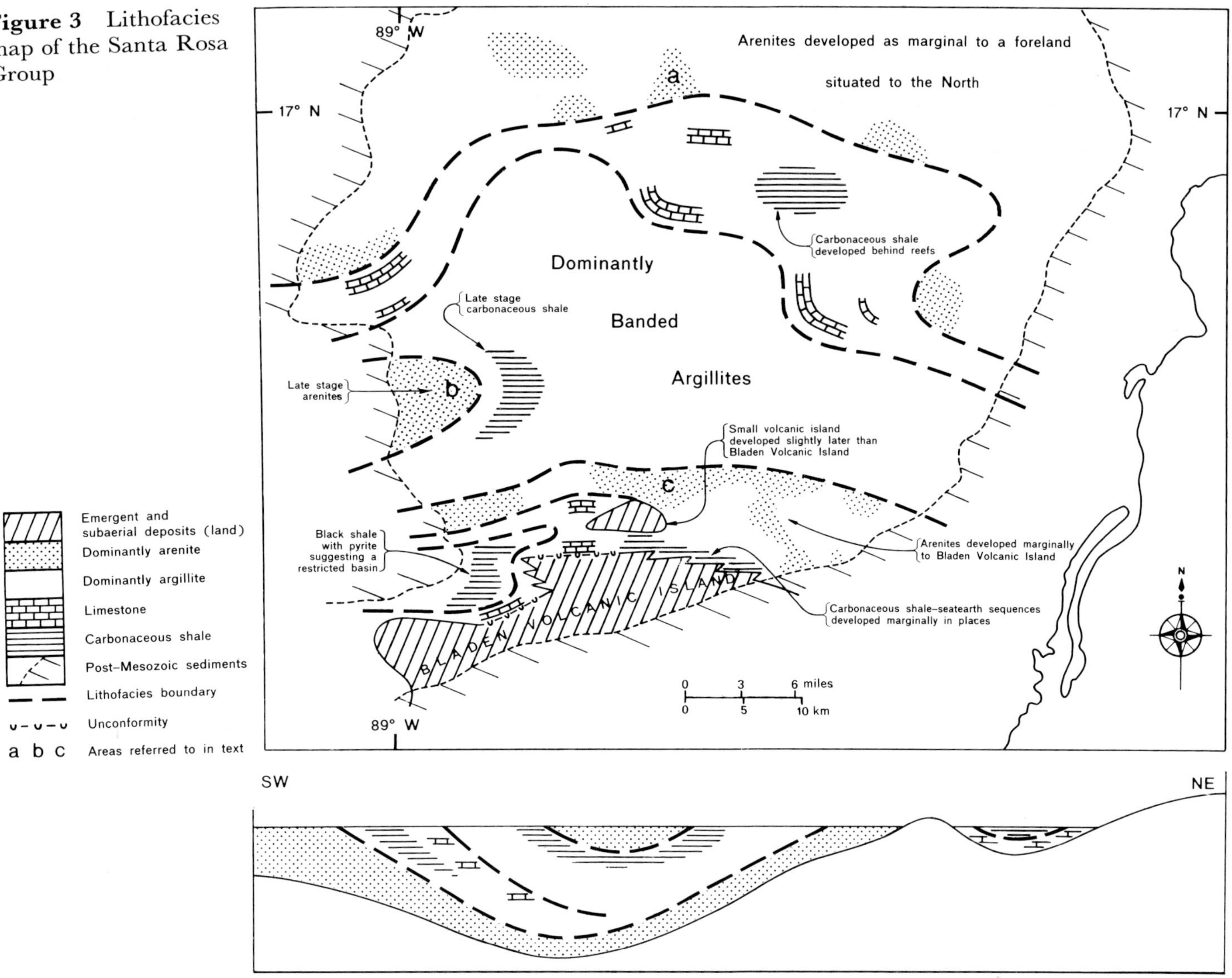

the base of the greywacke often shows fluting and other types of sole markings.

In a few areas such as in the tributaries of the Sibun River (17° 02′ N, 88° 41′ W), about 7 km north-east of Baldy Sibun, the lithology is dominated by a series of well-bedded, fine-grained grey quartzites (Plate 2.3 and 2.4). These are dense and composed of little other than fine granular quartz and a minute amount of sericitic mica.

The arenaceous rocks in general show few effects of the regional metamorphism. The matrices, however, show the development of sericite and chlorite and in the cleaner, better-sorted sandstones there has been a certain amount of welding and suturing of adjacent quartz grains. This last effect is best observed where sandstones are found close to the intrusive granites and have suffered a degree of thermal metamorphism.

Argillaceous rocks

Fine-grained rocks occupy by far the largest part of the area of the Maya Mountains. By their nature these rocks show the effects, in mineralogy and texture, of both the regional and thermal metamorphism to a greater degree than any of the arenaceous rocks. The argillites are, almost without exception, grey or black in colour, generally with fine sedimentary banding although in many localities the beds are massive. The more obviously banded deposits commonly exhibit fine, small-scale current bedding, with grading and some bottom scouring in the slightly coarser varieties. Locally ripple marks up to 2 cm amplitude have been developed.

In specimens coarse enough to allow mineral identification the bulk of the granular material is quartz set in a matrix in which sericite, chlorite and finely disseminated iron oxides are important constituents. The proportion of iron oxides determines the colour of the fresh rock and the red-brown colour of the clay into which the argillites are readily weathered.

A common feature of all the fine-grained rocks is a well-developed, close jointing pattern with two or three directions (Plate 3.1). Cleavage is commonly developed adjacent to the granites and main faults. The regional metamorphic

Plate 2 Photomicrographs of arenaceous rocks from the Santa Rosa Group

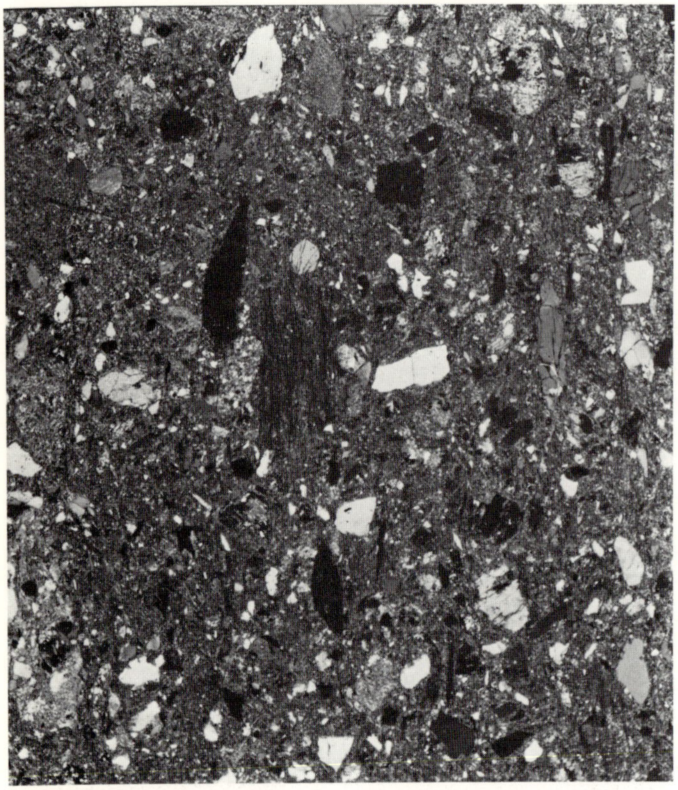

1 Greywacke from Trio Branch. Crossed polars (×12).

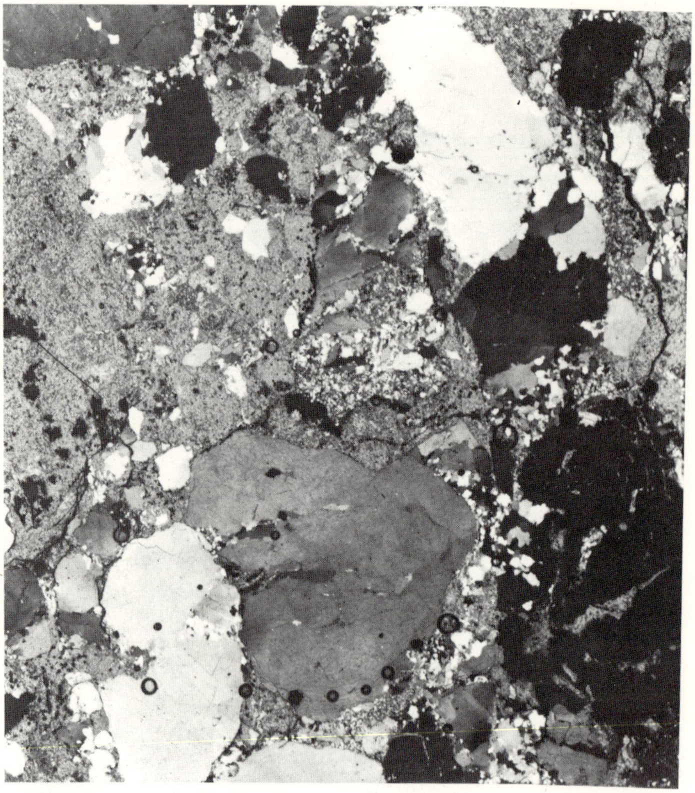

2 Subgreywacke from Macal River. Crossed polars (×6.5).

3 Quartzite from Mountain Pine Ridge. Crossed polars (×12).

4 Quartzite with grains showing sutured margins from Mountain Pine Ridge. Crossed polars (×12).

Plate 3
Rocks of the Santa Rosa Group

1 Typical well-cleaved mudstones, Mountain Pine Ridge.

2 Conglomerate with pebbles of sandstone, quartzite, phyllite and hornfels which forms an important part of the succession in the Baldy Beacon–Baldy Sibun area.

3 Limestone showing sedimentary banding, Cocoa Branch, Sittee River.

grade is responsible for the development of the chloritic and sericitic dust of the matrix which is generally very fine. Near the granites, however, local thermal effects are superimposed, resulting in the development of biotite, andalusite/chiastolite, staurolite and occasionally garnet, along with a general recrystallisation and not infrequently a marked schistosity.

Limestones

Dark limestone bands are associated with the argillaceous rocks. Their lateral extent is not known and they are generally thin although some thicknesses up to 30 m have been recorded (Dixon, 1956, p. 19). They have a characteristic honeycomb weathering on exposed surfaces. With few exceptions these bands are composed of the fragmentary remains of crinoids, some bivalves and, in a few instances from the Trio Branch (16° 34′ N, 88° 39′ W), both solitary and colonial corals. Dixon attached considerable importance to these crinoidal limestones which he regarded as being confined to and characteristic of the Macal Series (Dixon, 1956, p. 19). He attempted to use several of the limestones as marker horizons in order to arrive at an estimate of the thickness of the series in various parts of the region (Dixon, 1956, p. 18). Recent work has proved the existence of many more crinoidal limestone bands at various stratigraphic levels. This has led to the conclusion that their value for estimations of correlation and thickness is doubtful. Limestones of this type have been located not only in the northern part of the outcrop but also in several areas in the south, particularly in the upper reaches of the Trio Branch.

A larger unit of limestone has now been mapped in the headwaters of the Cocoa Branch (16° 51′ N, 88° 33′ W) of the Sittee River situated just north of the Cockscomb Granite and roughly central to the outcrop of the Santa Rosa Group. These beds (Plate 3.3), although noted by Forestry Department officers in the nineteen-fifties, and by geologists of D. R. Derry Ltd. in 1968, have never before been examined in detail. The main outcrop of these much-faulted sediments is indicated on the geological map from which they are seen to form a syncline plunging westwards (Figure 4). The limestone consists of a series of various lithological types, some composed of disoriented crinoidal fragments essentially similar to the smaller units found elsewhere while others are more massive and almost

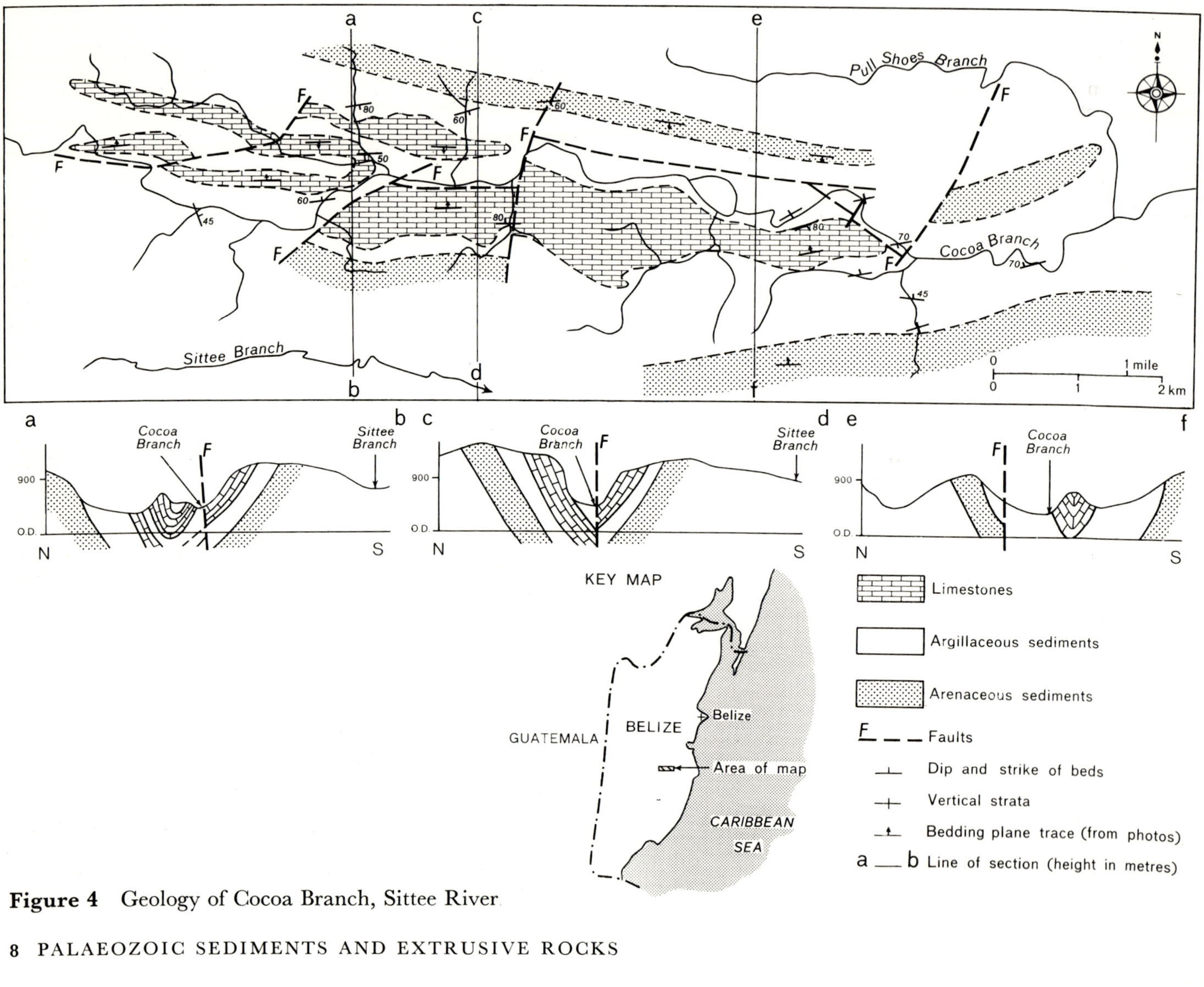

Figure 4 Geology of Cocoa Branch, Sittee River

entirely devoid of recognisable organic remains. Thin beds of shale and mudstone are also found between the limestones. Several of the individual limestone units contain grains of quartz (Plate 4.1.) In some similar clastic limestones the fragments are dark-coloured boulders and pebbles of crinoidal limestone set in a finer-grained recrystallised limestone. These rocks have been dated as probably Permian on the evidence of foraminifera (Plate 4.2 and 4.3). Also associated with these distinctive limestones are relatively well-sorted orthoquartzites which contain an appreciable proportion of carbonate in the matrix, indicating deposition in a carbonate-rich environment.

It is possible that these limestones are the equivalent of the Chochal Formation of Guatemala. Fossils obtained from the Chochal Formation include crinoids, fusilinids, algae, brachiopods and bryozoans ranging in age from Wolfcampian to Wordian (Blount, 1969). The assemblage is regarded as indicating that these rocks were deposited on a subsiding shallow shelf under conditions resembling those of the marginal back-reef facies of the Delaware Basin of west Texas.

BLADEN VOLCANIC MEMBER

A thick sequence of volcanic rocks outcrops along the southern margin of the Maya Mountains extending in a roughly east–west belt for about 50 km and varying in width from about 3 to 10 km. This was first mapped by Sapper (1889) and then by Ower (1928) who described it as an intrusive porphyry although he considered it to include flows and probably ash beds as well. Dixon (1956, p. 38) called this sequence the Bladen Porphyry and he considered it to be a laccolith.

The survey of 1969/1970 showed that the Bladen Porphyry is a complex of lavas, pyroclastic rocks and associated volcanic sediments (Hall and Bateson, 1972); and the rock group has consequently been renamed the Bladen Volcanic Member of the Santa Rosa Group (Bateson and Hall, 1971). The main outcrop of the Bladen Volcanic Member has six discrete areas of lavas surrounded by pyroclastic rocks and volcanic sediments. Two smaller areas of presumed lavas, which were not examined during the survey, occur in the extreme south-west of the Maya Mountains and appear to be isolated from the main outcrop of volcanics. One, forming Quartz Ridge, approximately 22.5 km south-west of Smokey River, was recorded as porphyry by Dixon, and the other, in the headwaters of Smokey River, was identified by the presence of porphyry in the float downstream and its outcrop delineated from the aerial photographs.

The lavas (Plate 5.1–5.4) are almost entirely rhyolitic and vary in texture from felsitic to coarsely porphyritic with feldspar and/or quartz phenocrysts up to 2 cm across. Flow-banding is commonly developed and autobrecciation is also frequently seen, particularly in the tops of individual flows and their margins. Spherulitic structures, in various degrees of development, are also found. The lavas are best developed in the area just north of the Bladen Branch (16° 31′ N, 88° 51′ W) where a series of flows form typical trap features on the hillside (Figure 5). Further north, in the area immediately to the south of the Trio Branch, individual flows are much more difficult to define, and flow-banding and autobrecciation are common. The associated sediments in this area are mainly reworked breccias. One volcanic cone has been identified in this area about 1.5 km to the south-south-east of Richardson Peak (16° 34′ N, 88° 46′ W). Its conical shape, rising about 100 m above the flat surface of one of the lavas, is clearly visible from the Southern Highway. No actual exposure of rock in situ was seen, but all the boulders and fragments on the cone are typically vent-breccia material. Further west, in the headwaters of the Bladen Brach and in Ramos Creek (16° 32′ N, 88° 52′ W), the lava flows are thinner and interbedded with tuffs, though the distinction in the field is made difficult by the effects of faulting and shearing.

Analyses of rocks from the Bladen Volcanic Member show them to be alkali rhyolites, and analyses 1 and 2, Table 1, compare closely with those quoted for typical sodic-potassic types, namely, 5 and 6. Analyses 3 and 4 are of rocks thought to have undergone secondary chemical alteration by metasomatism or metamorphism. No. 3 is described as a white laminated rock and is probably a rhyolitic tuff, and No. 4 as a sheared porphyry (M. N. Bass, written communication, 1969). Analyses of unaltered prophyry show that they are uniformly higher in total alkali than the granites of the Maya Mountains, analyses

Table 1 Analyses of rocks from the Bladen Volcanic Member, Belize

Sample	1	2	3	4	5	6
	%	%	%	%	%	%
SiO_2	73.51	71.39	79.32	71.61	73.20	72.78
TiO_2	0.18	0.48	0.07	0.53	0.16	0.19
Al_2O_3	13.71	13.63	11.88	13.43	10.75	12.86
Fe_2O_3	0.90	1.20	0.52	0.41	0.95	3.14
FeO	0.84	2.26	0.34	2.20	1.02	0.26
MnO	0.03	0.06	tr.	0.04	0.37	0.09
MgO	0.08	0.16	0.33	1.64	0.15	0.83
CaO	tr.	0.07	tr.	0.22	0.76	0.92
Na_2O	3.25	3.01	4.90	1.13	3.78	2.31
K_2O	5.96	5.91	1.44	6.52	4.20	4.14
H_2O 105°+	1.11	1.42	1.07	1.70	4.52	1.89
H_2O 105°−	0.26	0.22	0.23	0.14	0.18	1.05
P_2O_5	0.06	0.03	0.03	0.14	0.19	0.14
BaO	n.d.	n.d.	n.d.	n.d.	0.05	–
Other						0.08
Total	99.89	99.84	100.13	99.71	100.28	100.68

n.d., not determined
tr., trace

Samples 1 and 2. Porphyry from Bladen Creek.
Sample 3. (?)Rhyolitic tuff from Bladen Creek.
Sample 4. Sheared porphyry from north of Bladen Creek.
Sample 5. Sodic-potassic pitchstone, Arran, Scotland.
Sample 6. Rhyolite, Pentland Hills, Scotland.

Samples 1, 2, 3 and 4. Analyst: Tokiko Tiba (written communication, M. N. Bass, 1970).
Sample 5. Analyst: E. G. Radley (P. 234 in Geology of Arran. *Mem. Geol. Surv. G.B.*, 1928).
Sample 6. Analyst: W. Pollard (Geology of the Neighbourhood of Edinburgh, 2nd Ed., *Mem. Geol. Surv. G.B.*, 1910).

Plate 4 Photomicrographs of calcareous rocks from Cocoa Branch, Sittee River

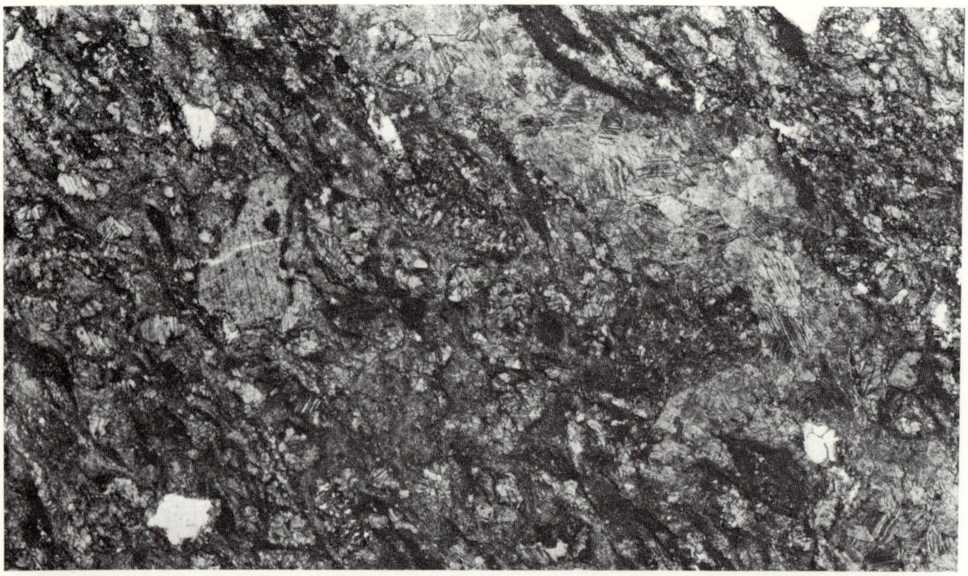

1 Limestone with detrital quartz grains. Plane-polarised light (×14).

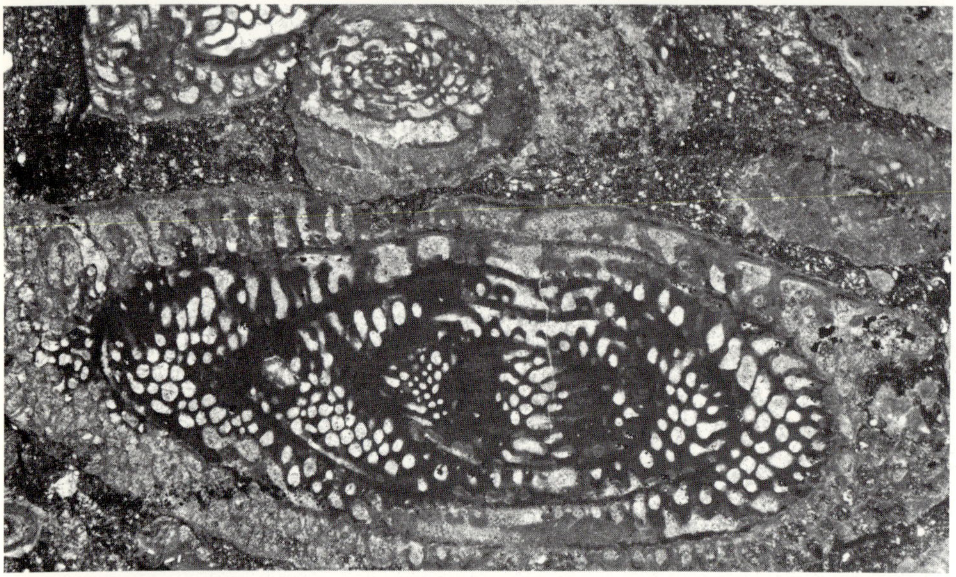

2 Fusulinid foraminifera. Crossed polars (×12.5).

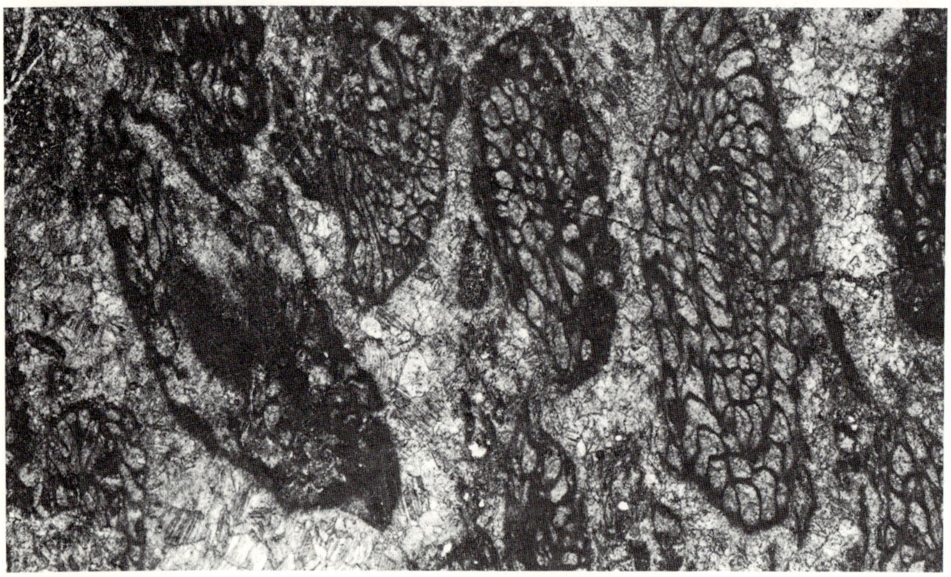

3 Fusulinid forminifera. Crossed polars (×12.5).

Plate 5 Photomicrographs of rocks from the Bladen Volcanic Member

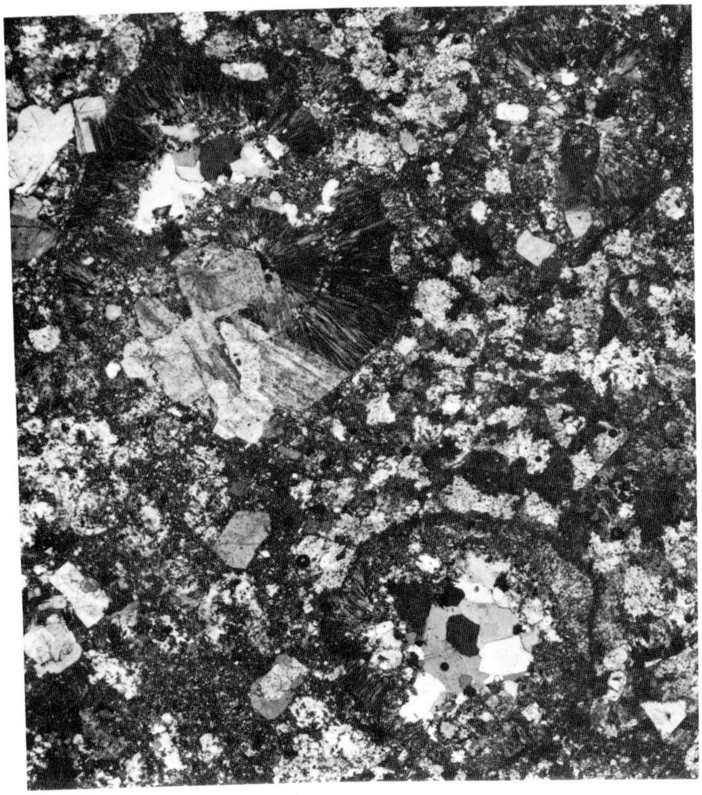

1 Spherulitic rhyolite from Richardson Creek, Bladen Branch. Crossed polars (×12).

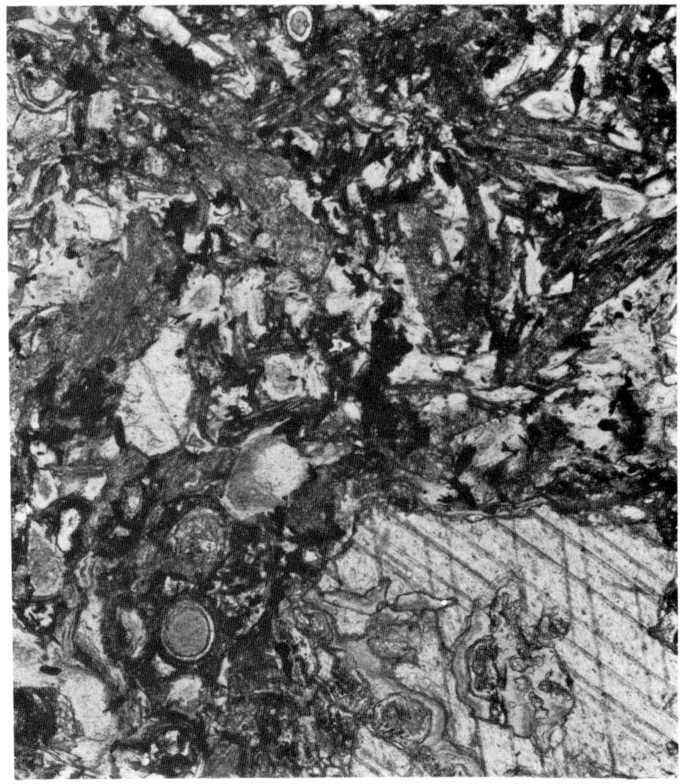

2 Pumice from Trio Branch. Crossed polars (×47).

3 Porphyritic rhyolite from Bladen Branch. Crossed polars (×17).

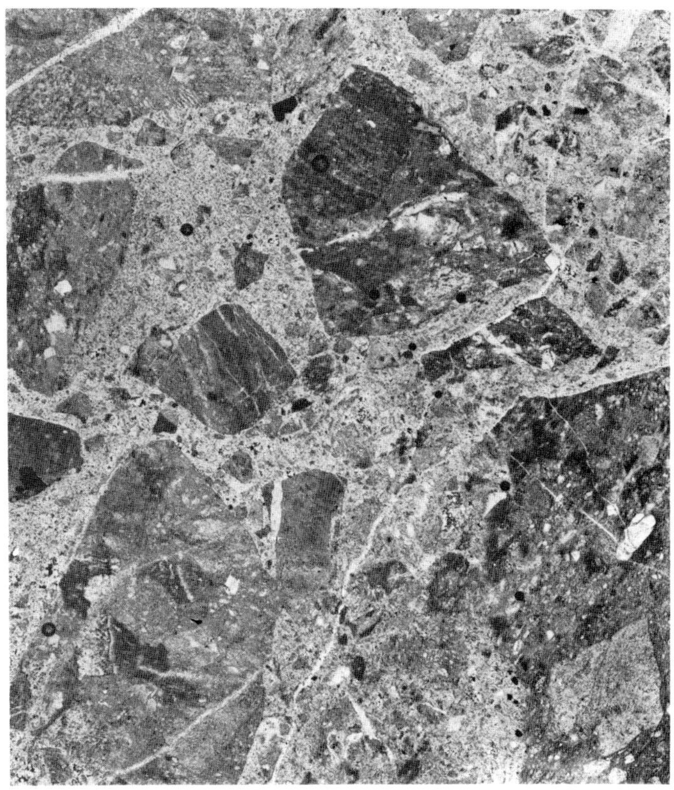

4 Autobrecciated rhyolite from Bladen Branch. Crossed polars (×13).

of which are shown in Table 2. Although there is considerable textual variation, and grain size ranges from wholly cryptocrystalline to fine-grained, with phenocrysts 2 cm or more in size (Dixon, 1956, p. 39), the chemical composition appears to be very similar in all areas examined. Where the rock is prophyritic the phenocrysts are feldspar and quartz, though the latter may occur alone. The feldspars are mainly alkaline orthoclase, sanidine and perthitic microcline, but plagioclase (up to An_{30}) sometimes occurs as an accessory. Mafic minerals are uncommon although biotite occurs occasionally, and irregular patches of chlorite probably represent altered mafics. Apart from iron ores and plagioclase, the more common accessories are zircon, epidote, sphene and apatite. Devitrified glass is very common and

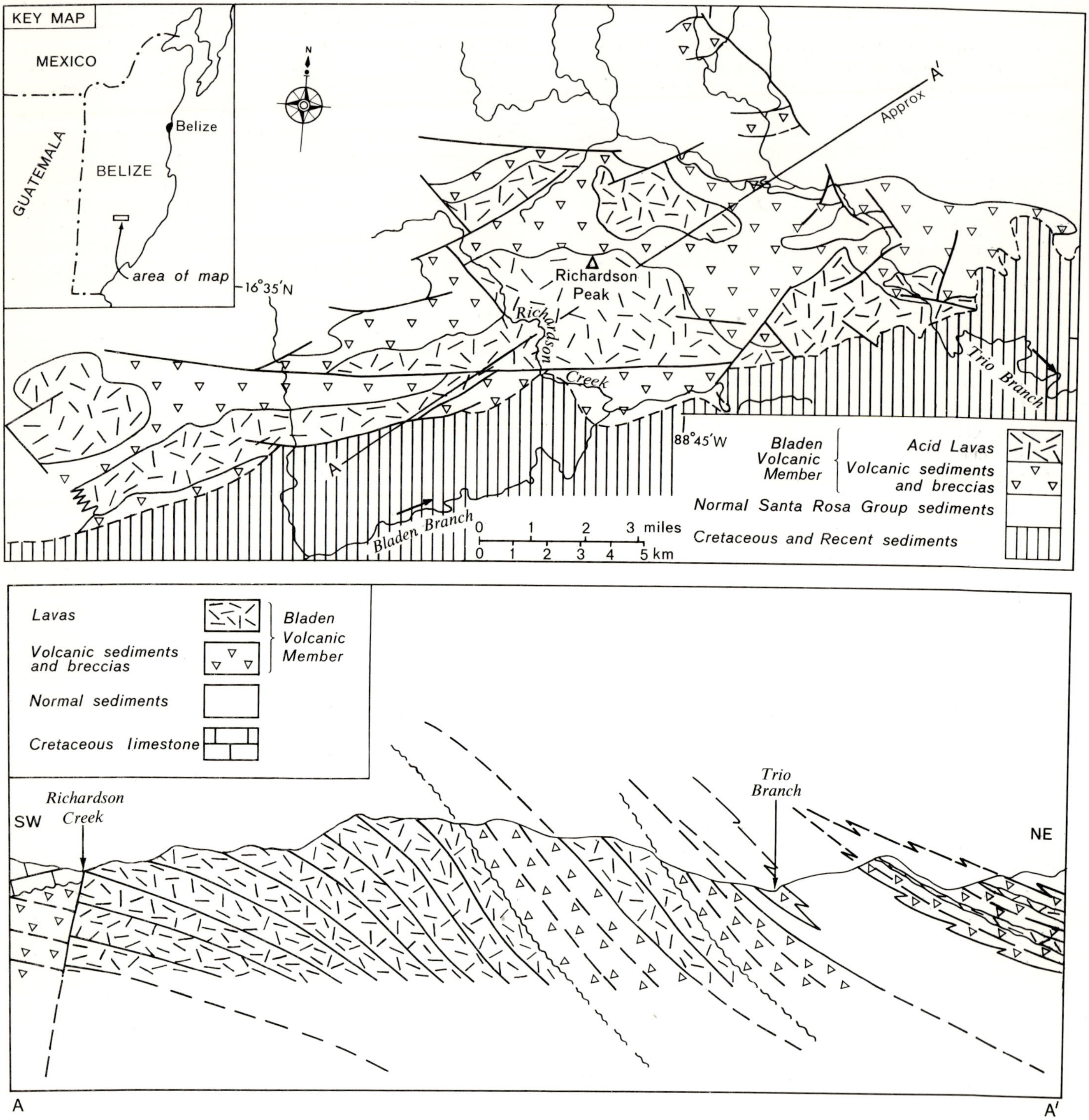

Figure 5 Geological map and cross-section of the Bladen Volcanic Member (after Hall and Bateson, 1972)

constitutes a large proportion of the rocks. Secondary minerals include sericite developed in varying degrees from feldspar, chlorite (often associated with iron ores), and less commonly, epidote which may also be associated with iron ores.

Samples were taken for more detailed petrographic examination from a flow, about 60 m thick, in the Richardson Peak area, north of the Bladen Branch. The top of the flow is a typically autobrecciated rhyolite with angular fragments in a fine-grained groundmass. The fragments are composed entirely of devitrified glass and the groundmass consists of irregular areas of devitrified glass, areas of quartz and slightly sericitised laths of potash feldspar. The centre of the flow is a devitrified porphyritic rhyolite. In this the devitrified groundmass consists of quartz and sericite, the latter probably derived from potash feldspar, and in places shows incompletely developed spherulitic texture. The phenocrysts consist of quartz, microcline-perthite formed by the exsolution of sanidine, and oligoclase. The oligoclase phenocrysts are smaller and make up much less than one third of the total feldspar in the rock. There are also occasional small epidotes (associated with some opaque minerals) and rare zircons. The base of the flow is a very fine felsite with a cryptocrystalline groundmass of devitrified glass and scattered small subhedral phenocrysts of largely sericitised alkali feldspars and occasional quartz. Irregular patches of chlorite associated with opaque minerals, probably represent altered mafic minerals, and occasional zircons occur. Sericitisation is more completely developed towards the base of the flow, and heavier accessory minerals also increase downwards. As this flow is about 60 m thick, cooling was probably sufficiently slow at the centre to allow a certain amount of settling of the heavier minerals to take place despite the high viscosity of the lava. Exsolution of the sanadine phenocrysts to microcline-perthite could also have taken place during this period.

About 7 km north-east of Richardson Peak (16° 38′ N, 88° 44′ W), in a southward-flowing tributary of the Trio Branch a small area has been identified as containing volcanic breccias, sandstone, a few thin lava flows and some pyroclasts. They are interbedded and folded with the normal shallow water sediments found round the Bladen Volcanic island. They occur on the north and south flanks of a syncline about 4 km across. Their lateral extent is about 3 km. Towards the centre of the area and upwards in the succession, the normal sediments become interbedded with reworked volcanic sandstones which often show slump structures. These give way to completely unsorted, unstratified breccias which appear to have been deposited in a subaerial environment, but in some areas are unconformably overlain by volcanic sandstones and normal sediments. Some reworked volcanic sandstones, overlain by sequences of sandstone, seat earth and carbonaceous shale, are interbedded with the breccias and may indicate that the volcanic island was intermittently uplifted. The associated lavas are thin, typically vesicular, sometimes having a ropy structure, but do not show any evidence of submarine eruption. One of the lavas has been found to be an oligoclase-bearing andesite. It has a felted texture consisting of oligoclase–andesine laths about 0.2 mm long and a groundmass of oligoclase–andesine microliths. There are a few pyroxenes with pseudomorphs of limonitic material among the laths, and a little pyroxene and rare muscovite in the groundmass. Quartz and potash feldspar are notably absent. There are also numerous vesicles many of which are empty and the rest are partly or completely filled with penninite. Small blocks of pumice, with a very similar composition to the lavas, occasionally occur in the volcanic sediments. This relatively small area is interpreted as being a short-lived volcanic island completely separated from the main Bladen Volcanic island and developing a little later than the main period of vulcanicity.

Surrounding the various lava masses of the Bladen Volcanic Member is a series of pyroclastic rocks and volcanic sediments. Vitric and lithic tuffs are found interbedded with the lavas but, around the margins, breccias are found which consist almost entirely of felsitic fragments. These were probably formed by penecontemporaneous subaerial erosion of the lavas. The breccias are commonly reworked to give conglomerates and sandstones composed entirely of volcanic material; in places the breccias can be seen to pass laterally into the sandstones. The volcanic sandstones are interdigitated with normal marine sediments. Thus it can be deduced that the lavas are more or less contemporary with the normal sediments and form an integral part of the succession. Along the north-west margin of the Bladen Volcanic island normal sediments unconformably overlie the volcanic sandstones and may indicate overlap in that area. Rather meagre palaeontological evidence indicates that the sediments in this area are not older than Upper Pennsylvanian and are probably mainly Permian; this is consistent with an absolute age of 300 Ma given for the lavas (Anon., 1968). The identification of the lavas as an integral part of the succession and the establishment of their age is of considerable significance in a regional context, since there are at present very few published data on volcanics of similar age from other parts of Central America.

THREE

Intrusive igneous rocks

The intrusive igneous rocks of the Maya Mountains comprise granites, porphyries and dyke rocks. The granites are the most extensive. They crop out in the Mountain Pine Ridge (16° 59′ N, 88° 56′ W) at various places close to the northernmost boundary of the Santa Rosa Group, and in the rivers and streams of the Hummingbird (17° 00′ N, 88° 28′ W) and Mullins River (17° 03′ N, 88° 25′ W) areas, and occupy a large basin immediately to the south of the Cockscomb Range of mountains (16° 49′ N, 88° 37′ W). The porphyries occur in two localities to the north-west of Baldy Beacon and as a dyke in the north of the Mountain Pine Ridge area at Slate Creek (17° 03′ N, 88° 57′ W). The relationship between the granites and the porphyries is uncertain. The third group was intruded at a later date; it includes a variety of aplitic dykes, rare pegmatites and two occurrences of dolerite dykes.

GRANITES

The granites of the Maya Mountains occur in three areas: the Mountain Pine Ridge area (Plate 6); the Cockscomb Range (16° 49′ N, 88° 37′ W), including the Sapote (16° 39′ N, 88° 35′ W) granite; and Hummingbird–Mullins River (17° 01′ N, 88° 26′ W) in the north-eastern quadrant of the area.

All the granites are mica-bearing and most of them have a higher proportion of biotite; they are generally leucocratic. Differences in the chemistry of the various granites are indicated in Table 2. Features have been noted in the recently completed geochemical reconnaissance of the area which showed in particular a noticeably low boron content in samples derived from the Mountain Pine Granite in comparison with stream sediments from the other granite masses.

Mountain Pine Granite

This granite mass is generally characterised by being medium-grained, nonporphyritic, nonfoliated and leucocratic. It exhibits, however, a number of different mineralogical facies (Plate 7) which are thought to indicate several phases of intrusion.

Macroscopic and microscopic examination of samples from this granite shows the main mass to be essentially a two-mica granite (Plate 7.1) in which the biotite is in excess of the small percentage of muscovite. The biotite is a green, pleochroic variety which occurs in clumps of small crystals. The crystal margins are commonly chloritised with pleochroic haloes. The white mica is usually associated with the biotite and, in some places, may be biotite completely leached. The feldspars are large anhedral crystals of perthite and microperthite with a small percentage of plagioclase, which occurs as small discrete crystals showing carlsbad-albite and simple albite twinning with a composition close to the oligoclase–andesine boundary. Characteristically the plagioclase, unlike the potash feldspar, shows considerable alteration to small flakes of sericite mica; in some instances alteration is so far advanced that all trace of twinning has been obliterated leaving only the outline of the original plagioclase. Quartz occurs as pools of slightly strained grains and also as smaller grains in the matrix. It is estimated that the quartz is in excess of 10 per cent of the rock and the potash feldspar is much more abundant than the plagioclase. Apatite and magnetite are the commonest accessory minerals.

Analyses of samples from the Mountain Pine Granite are given in columns 1 and 2 of Table 2. The composition is similar to that of a potassic granite from Dartmoor (analysis 5 in Table 2).

The Mountain Pine Granite is generally uniform, but there are local variations in texture and mineralogy. Foliation is developed in some places, noticeably in the area north of Privassion Creek (17° 02′ N, 88° 58′ W) where faulting within the granite—possibly associated with the Northern Boundary Fault zone (17° 05′ N, 88° 43′ W)—has caused shearing. This has been accompanied by the development of chlorite and by fracturing and rounding of

Plate 6 View of a typical granite outcrop, Mountain Pine Ridge

Plate 7 Photomicrographs of granitic rocks

1 Mountain Pine Granite. Crossed polars (×13).

2 Hummingbird Granite. Crossed polars (×7).

3 Sapote Granite. Crossed polars (×13).

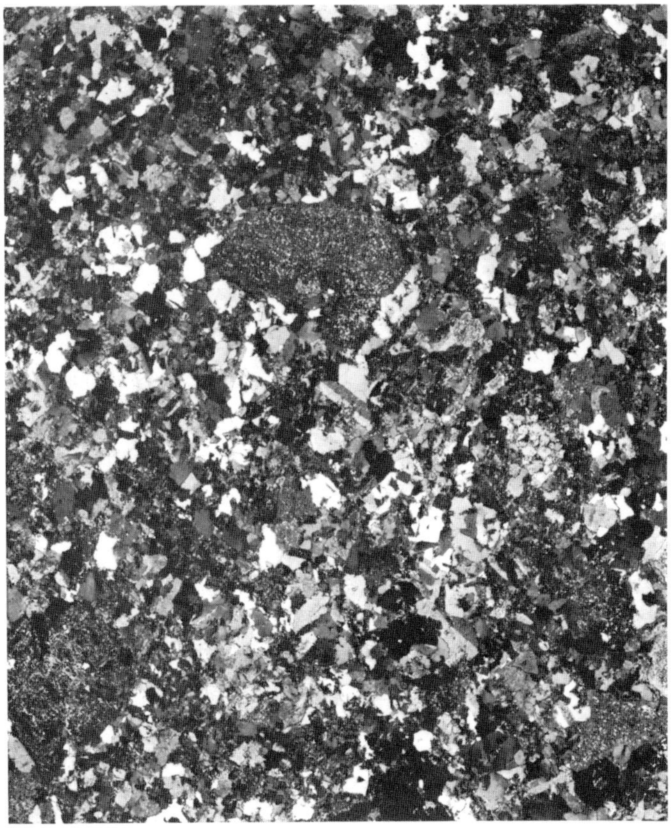

4 Aplite. Crossed polars (×12).

the feldspars particularly in the Barton Creek area (17° 05′ N, 88° 55′ W) where phenocrysts of feldspar are characteristic. In some places the balance between the biotite and muscovite content changes so that locally the rock becomes a muscovite-granite. Tourmaline was recorded from one small area north of Guacamallo (16° 52′ N, 89° 03′ W). During this investigation no attempt was made to delimit the various mineralogical facies of this granite. Radiometric age determinations on rock samples taken from this batholith fall close to an isochron giving an age of 320 ± 10 Ma—a Late Carboniferous date that is consistent with the field data.

Cockscomb–Sapote Granite

The Cockscomb–Sapote Granite has many characteristics in common with the Mountain Pine Granite. It is, for example, a two-mica granite (Plate 7.3). The biotite, a dark-brown pleochroic variety, is in excess of the muscovite, which occurs as scattered cleavage fragments. The biotite, in some of the thin sections examined, shows an abundance of minute needle-like inclusions, rutile, which are regularly arranged parallel to the edges of hexagonal crystals and impart a false three-directional cleavage particularly in basal sections. The feldspars are a mixture of perthite, some microcline and orthoclase with smaller crystals (often showing carlsbad/albite twinning) of plagioclase with a composition near the oligoclase–andesine boundary. Many of the plagioclase crystals have a clear outer rim of albite. Alteration of the plagioclase is common. Many crystals are identifiable only in outline, the mineral itself being completely replaced by a mass of white mica. A poikilitic relationship almost invariably exists between the plagioclase and the microcline and perthite.

Quartz occurs in large pools and is sufficiently abundant for the rock to be classified as a granite. Accessory and alteration minerals are associated with the clusters of biotite. Apatite forms euhedral crystals and zircons occur at the centre of many of the pleochroic haloes in the biotite. In less fresh samples, particularly those derived from localities close to faults and crush zones, the biotite shows alteration to chloritic minerals with associated clusters of brown epidote. Clinozoisite forms a vein in one sample and in thin section is locally associated with clumps of mafic minerals. Monazite is a rare accessory mineral. Its occurrence was confirmed in heavy mineral concentrates, particularly from the Sapote Granite.

Hummingbird–Mullins River Granite

The Hummingbird–Mullins River Granite (Plate 7.2) is similar in composition to the other granites but locally the assimilation of country rock has resulted in a more basic marginal facies. Quartz occurs as groups of interstitial grains rather than the large pools typical of the Mountain Pine and Cockscomb granites and makes up a smaller proportion of the rock than is general elsewhere. Normal accessory minerals are apatite, zircon and magnetite. The analyses of two samples of this rock are listed in Table 2. The granite is appreciably lower in silica and total alkalies, but correspondingly richer in alumina and total iron than the Mountain Pine granite. An analysis of a granodiorite is included in Table 2 for comparison.

The rock becomes more granodioritic or dioritic near the intrusive margin. Rare garnet in the granite and in the nearby metamorphosed country rock is evidence of contamination by the original sediments. Dixon recorded general basification of the normal granite in the Pomona and False Creek areas (16° 58′ N, 88° 23′ W) and in the exposures in the Mullins River (17° 03′ N, 88° 25′ W) (Dixon, 1956, p. 36).

Table 2 Analyses of granitic rocks from Belize

Sample	1	2	3	4	5	6
	%	%	%	%	%	%
SiO_2	75.58	75.05	65.19	66.44	73.66	66.91
TiO_2	0.25	0.24	0.72	0.61	0.16	0.67
Al_2O_3	12.59	12.29	16.06	15.21	13.81	15.09
Fe_2O_3	0.15	0.65	0.22	0.54	0.21	1.70
FeO	1.49	1.39	4.02	3.60	1.51	1.95
MnO	0.03	0.03	0.11	0.11	0.06	0.18
MgO	0.12	0.47	2.11	1.88	0.45	2.02
CaO	0.59	tr.	3.61	1.66	0.67	3.27
Na_2O	2.76	2.78	3.14	3.46	2.89	4.16
K_2O	5.20	5.54	2.25	3.01	5.02	3.16
H_2O 105°+	0.76	1.02	2.32	2.77	1.25	0.14
H_2O 105°−	0.16	0.17	0.30	0.31	0.41	0.56
P_2O	0.05	0.04	0.13	0.09	0.24	0.17
BaO	n.d.	n.d.	n.d.	n.d.	n.d.	n.d.
Total	99.73	99.67	100.18	99.69	100.35	100.03

n.d., not determined
tr., trace

Samples 1 and 2. Granites, Mountain Pine Ridge.
Sample 3. Granodiorite, Hummingbird Gap.
Sample 4. Sodic granite, Hummingbird Highway.
Sample 5. Potassic granite, Haytor East Quarry, Dartmoor.
Sample 6. Granodiorite, Moor of Rannoch.

Samples 1, 2, 3 and 4. Analyst: Tokiko Tiba (written communication, M. N. Bass, 1970).
Sample 5. Analyst: H. F. Harwood (*Mineral. Mag.*, Vol. 20, p. 41, 1923).
Sample 6. Analyst: E. G. Radley (The geology of Corrour and the Moor of Rannoch. *Mem. Geol. Surv. G.B.*, 1923).

PORPHYRIES

Fine-grained acidic porphyries (Plate 8) crop out at two localities to the north-west of Baldy Beacon (17° 01′ N, 85° 45′ W). Their composition is granitic, comprising essentially a fine-grained quartzo-feldspathic matrix in which are embedded phenocrysts (up to 3 mm long) of quartz, perthite, and sericitised plagioclase feldspar. The quartz phenocrysts are commonly corroded. Small areas of chlorite occur associated with granular magnetite and euhedral crystals of zircon and apatite; these probably replace original biotite.

The spatial relationship between these quartz-feldspar porphyries and the Mountain Pine Granite nearby is unknown. The porphyries probably post-date the arenaceous facies of the Santa Rosa Group because the sediments show evidence of thermal metamorphism. A small dyke of

Plate 8 Photomicrograph of porphyry cutting the Mountain Pine Granite. Crossed polars (×13)

Plate 9 Photomicrograph of dolerite cutting the Mountain Pine Granite. Crossed polars (×18)

similar quartz-feldspar porphyry cuts argillites at Slate Creek (17° 03′ N, 88° 57′ W) which have been thermally metamorphosed by the Mountain Pine Granite. This relationship would seem to indicate that this porphyry, and possibly those of Baldy Sibun, are younger than the granite. Similar igneous rock is reported by Dixon (1956, p. 37) from close to Little Vacqueros Creek (17° 03′ N, 88° 58′ W) within the Mountain Pine Granite, but here also the relationship with the granite is not clear.

DYKE ROCKS

A few dykes of aplite, pegmatite and dolerite occur in the area of the Maya Mountains.

The aplite dykes (Plate 7.4) consist almost entirely of granular quartz and feldspar in approximately equal proportions, with very minor quantities of muscovite, biotite and chlorite. The few examples known all appear to be intrusive into the Mountain Pine Granite and are thus younger. The pegmatites are similarly uncommon and have been observed only in the Mountain Cow Creek (16° 57′ N, 88° 26′ W) as small dykes and pods. Float, rich in tourmaline and mica, has been found in a tributary of the Bladen Branch which cuts through the Sapote Granite (16° 39′ N, 88° 36′ W). This is thought to indicate the presence of pegmatitic material in this area.

Basic igneous activity is insignificant in this area. A medium-grained dolerite dyke forms a discontinuous train of boulders, striking N060°, north-east of San Luis (16° 45′ N, 89° 01′ N) in the Augustine–Mountain Pine Ridge area (Plate 9). One small dolerite dyke cuts granite in the headwaters of the Silk Grass Creek (16° 55′ N, 88° 20′ W); another occurs in the upper Mullins River (17° 03′ N, 88° 25′ W).

G. H. Wilson, in his notes in 1885, recorded the occurrence of serpentine and nephrite as float in the Trio Branch (16° 34′ N, 88° 39′ W). His localities could not be positively identified during the recent survey but no evidence for the occurrence of these rock types was found. The available chemical and petrological data indicate that the occurrence of such rock types is very unlikely in this environment.

FOUR

Post-Permian sediments

The Palaeozoic rocks of the Maya Mountains are overstepped in the west by a thick series of limestones and dolomites with some calcareous shales, siltstones and sandstones. Similar sediments are found in the areas to the north and south of the Palaeozoic rocks of the Maya Mountains with which they are known to be in faulted contact. Various authors have examined these deposits in the area to the north of the Maya Mountains from which sufficient fossil evidence has been obtained to allow some dating of the rocks and correlation with Guatemala.

Dunlop (1920) gave the age of the limestones as Lower Cretaceous to Eocene but did not indicate the source and nature of his evidence. Ower (1928) reported the occurrence of Eocene fossils in these rocks and correlated them with the Rio Dulce Limestone of Guatemala which Schuchert (1935) stated to be of Lower Miocene age. In more recent years Flores (1952) examined the sediments of the northern part of British Honduras on behalf of the Bahamas Exploration Co. Ltd. and his work remains a standard authority on these rocks. He showed the limestones in the north to be Upper Cretaceous and Middle Eocene but obtained no evidence to suggest the presence of any deposits of Oligocene age. He did, however, suggest that some of the limestone conglomerates found between El Cayo, near the international border with Guatemala, and Roaring Creek may represent the talus deposits formed during an Oligocene emergence. The Eocene to the south of the Maya Mountains is represented by the shales, mudstones and calcareous sands of Dixon's Toledo Series. The evidence for the Cretaceous age of the limestone in this southern area is based upon the occurrence of rudistid fragments found and recorded by Willis (1912), Reed (1921), Dickerson and Weisbord (1931), and Sapper (1937), and upon some foraminiferal limestones a few miles to the west of Punta Gorda which contain identifiable late Cretaceous foraminifera (Dixon, 1956).

The limestones to the north of, and resting on, the Maya block have undergone intense recrystallisation which has destroyed the majority of the fossil remains. These limestones are generally massive and associated with well-bedded dolomites.

The succession in the Caves Branch (17° 05′ N, 88° 44′ N)–Margaret's Creek (17° 06′ N, 88° 37′ W) area was determined by Flores (1952). He named the lower arenaceous series of rocks the Margaret Creek Formation which has a maximum thickness of 60 m. Fossils, other than lignitic and pyritic wood fragments, are very rare in these sediments. This is characteristic of the close-to-shore environment in which they were deposited. The Barton Creek Formation (17° 05′ N, 88° 55′ W) of limestones and dolomites, some 600 m thick, lies above the Margaret Creek Formation. Palaeontological evidence, although scarce, is sufficient to date these rocks as Upper Cretaceous. By analogy with these limestones the massive-bedded deposits, which extend westwards from Augustine (16° 58′ N, 89° 00′ W) and form the Vaca Plateau, are tentatively regarded as of probable Upper Cretaceous age. No evidence has been recorded, however, which indicates the presence of shallow estuarine deposits of the Barton Creek type on this part of the Maya Mountain block.

No deposits of Tertiary age have been mapped on the Maya Mountain block. The region was probably an area of emergence throughout post-Cretaceous times.

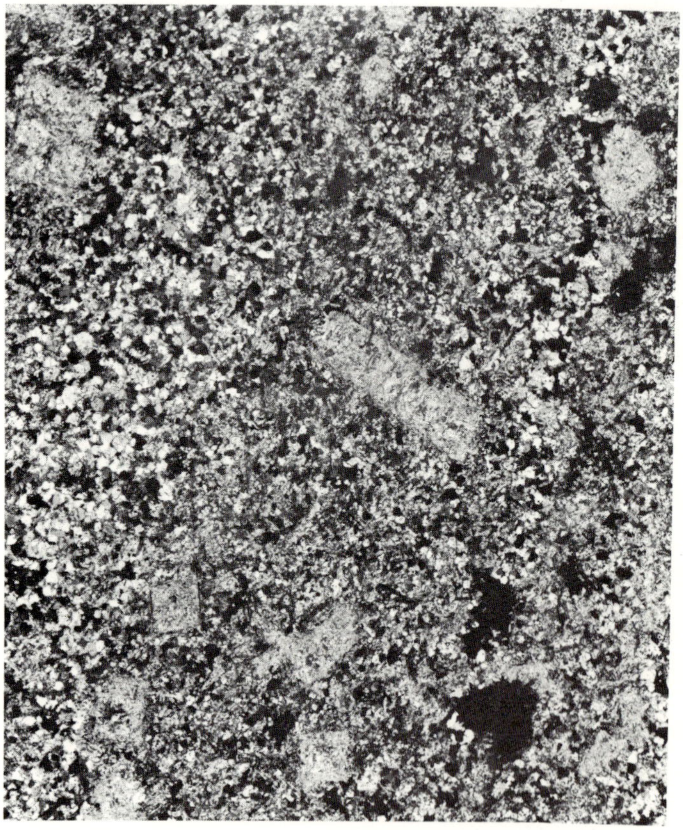

Plate 10 Hornfels with altered andalusite, from the metamorphic aureole of the Hummingbird Granite. Crossed polars (×17).

FIVE

Metamorphism

The sediments of the Santa Rosa Group have undergone widespread low-grade regional metamorphism throughout the area, but the younger Cretaceous sediments and the granites are unaffected. This metamorphism is identifiable in almost all the sedimentary rock types of the Santa Rosa Group and is normally exhibited by the presence of sericitic mica, chlorite and iron oxides.

The sericitic mica as minute laths with amorphous chlorite forms a significant part of the fine matrix of the sediments. Even in the coarser sediments the interstices between the quartz grains are packed with chlorite and sericite. The flakes of mica tend to be larger in the more argillaceous rocks and commonly have a preferred orientation producing a phyllitic cleavage. The chlorite occurs characteristically as irregular, pale-green, slightly pleochroic amorphous masses. The low grade of the regional metamorphism is further indicated by the fact that the quartz grains in all the sediments examined are still identifiable as clastic grains with little or no recrystallisation.

Plate 11 Hornfels with staurolite, from the metamorphic aureole of the Hummingbird Granite. Plane-polarised light ($\times 65$).

The occasional detrital zircons and tourmalines also retain their clastic characteristics. Iron oxide dust, mainly hematite, is scattered throughout many of the rocks, but in those samples that have undergone some degree of pressure the iron oxides show a tendency to alignment along the incipient cleavage traces. In the better sorted and coarser sandstones, the blades of sericite, with smaller amounts of chlorite and iron, are moulded around the configuration of the clastic grains in the sediment. In the limestones of the Santa Rosa Group the low-grade regional metamorphism gives rise to recrystallisation of the calcite with characteristic interlocking of the new grains. Those limestones with an appreciable quantity of clastic quartz also show these effects, and in the varieties that had an appreciable original clay fraction the formation of small laths of sericite is common.

Near the granites the sediments have also been subjected to thermal metamorphism which rarely exceeds the andalusite grade. The commonest of the metamorphic minerals formed within the aureole is andalusite (or chiastolite) (Plate 10). This mineral is quite commonly found in rocks of dominantly argillaceous composition in which crystals up to 15 cm have been observed. The occurrence of andalusite is less common in the more arenaceous beds. The fabric of such rocks has been completely recrystallised to produce a hornfels. The matrices contain sericite, chlorite and iron dust while the andalusite and/or chiastolite remain only as a felted mat of white-mica flakes. Sediments thermally metamorphosed to the andalusite grade are very common around the margins of the granites of the eastern part of the Maya Mountains but seem to be less well developed around the Mountain Pine granite complex. In some areas, notably in the sediments in the Silk Grass Creek (16° 55′ N, 88° 23′ W) watershed, garnets are common. Minute metamorphic garnets have also been found in a narrow aureole in the tributaries of the South Branch of the North Stann Creek (17° 00′ N, 88° 32′ W). Staurolite occurs in a fine-grained sandstone from the Silk Grass Creek (Plate 11).

Many of the regionally metamorphosed sediments, particularly those in the finer sand grades, contain tourmalines of detrital origin. Some rocks in the thermal aureole contain pale brown, pleochroic, euhedral tourmaline apparently formed contemporaneously with the other metamorphic minerals. It is thought that this tourmaline may represent the only real evidence of boron pneumatolysis of the metasediments, as the recrystallisation of detrital tourmalines under these conditions of metamorphism would probably have resulted in the tourmalines having a distinct cored or zoned appearance.

Locally the sediments have also been altered by the shearing effects of faults in which the region abounds. These effects can be seen in all types of rocks, the granites of the lower parts of the Sibun River and adjacent streams commonly showing marked cataclasis and shearing whilst the finer-grained sediments are converted into highly lustrous phyllites and schists.

SIX

Structure

The first modern study of Belize led Ower (1928, p. 507) to suggest that the Maya Mountains formed a geanticline with adjoining synclines to the north and south now filled with limestones. After a more comprehensive study, Dixon (1956, p. 43) considered the Maya Mountains to be an uplifted fault block which had been subjected to several orogenies. The Santa Rosa Group comprises his Maya and Macal Series, a subdivision erected partly on the basis that the former had been subjected to considerably more intense folding than the latter, and that the Maya Series had been affected by low-grade regional metamorphism whereas the Macal Series had not. The geological sections which accompany Dixon's report indicate that he considered the Macal Series to form a broad shallow syncline unconformably overlying the tightly folded Maya Series, which, together with the Bladen Porphyry, forms a structural high in the Trio Branch area. In their study of the macrostructure of a small area in the north-west of the Maya Mountains, Kesler and others (1971) found no evidence to support this subdivision, and the present survey has shown that it is not valid on either structural or stratigraphical grounds (Bateson and Hall, 1971).

It has been confirmed during the recent work that the overall structure of the Maya Mountains is an uplifted fault block consisting of a synclinorium trending east-north-east and plunging to the west at about 10°. The area is clearly bordered on the north and south by major faults, evidence for which can be seen in the field (Figure 6). The eastern boundary is probably also formed by a major fault striking north-north-east, parallel to the present coastline. The structure of the western boundary of the fault block is completely obscured by gently folded Cretaceous–Eocene limestones which lie unconformably on the Santa Rosa Group. This simple regional pattern is complicated by the intrusion of several granitic masses arranged in a semicircular arc, and by the presence of several faults, some of which have a considerable throw. No evidence for a structural high in the Trio Branch area was found (cf. Dixon, 1956). All the lava flows and the bulk of the associated sediments of the Bladen Volcanic Member dip northward, but a few southerly dips were recorded in metasediments along the southern margin of the area. These may be the southern limb of a small anticline within the main synclinorium, but are probably more likely to be related to drag folding associated with the Southern Boundary Fault.

FOLDING

The majority of bedding planes throughout the area dip steeply, and it is therefore probable that the sediments of the Santa Rosa Group are tightly folded. Similar steeply inclined bedding planes occur in the dominantly arenaceous sediments in the Baldy Beacon–Baldy Sibun area (17° 00′ N, 88°45′ W)—the type area of Dixon's Macal Series (Figure 7). It is possible to identify many closely spaced axial planes of folds within these sediments on aerial photographs (J. A. E. Allum, oral communication, 1971). These observations of steep folding throughout the whole area would appear to refute Dixon's concept of two distinct and differently deformed sedimentary series.

A stereographic projection of poles to bedding (Figure 8A) shows that maximum concentration occurs along a great circle, indicating regular folding about parallel axes, which is consistent with the existence of a major synclinorium shown by field mapping. The pole of this great circle gives an azimuth of about N070° and an axial plunge to the west of about 10°. The concentration of poles near the primitive indicates that the limbs of the folds are steeply inclined. However, in the area to the north of the Bladen Volcanic Member, drained by the Trio Branch and its tributaries (16° 38′ N, 88° 47′ W), the folding is much more open and, of a total of 25 readings with dips of less than 30°, 92 per cent came from this area. This concentration of abnormally low dips is interpreted as being caused by a local buttressing due to the subsurface extension of the thick pile of volcanics which, in this area, lies at no great depth under

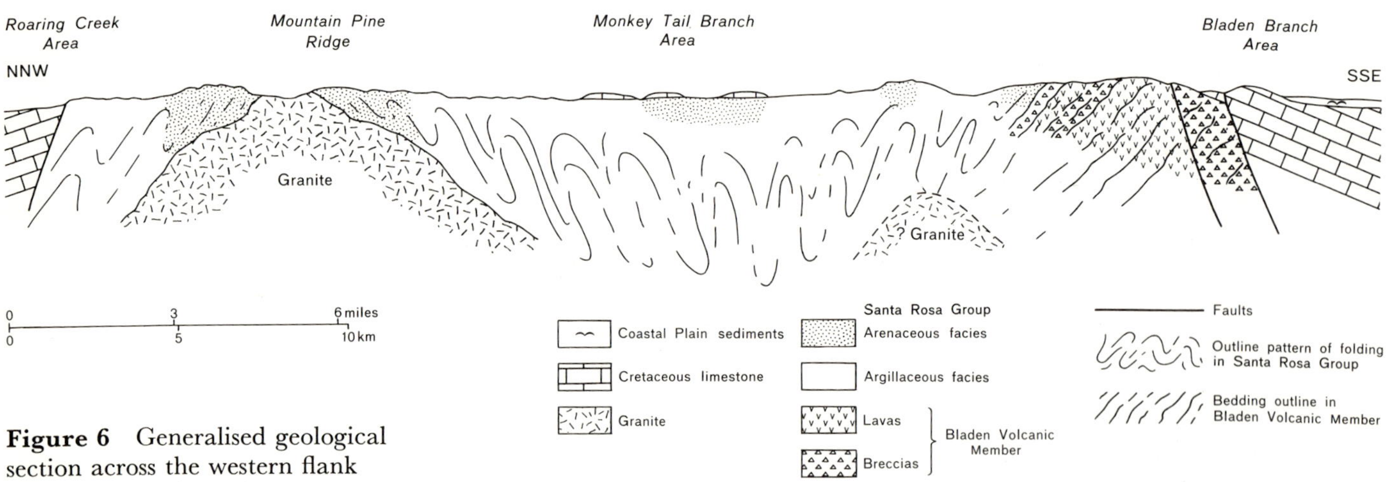

Figure 6 Generalised geological section across the western flank of the Maya Mountains

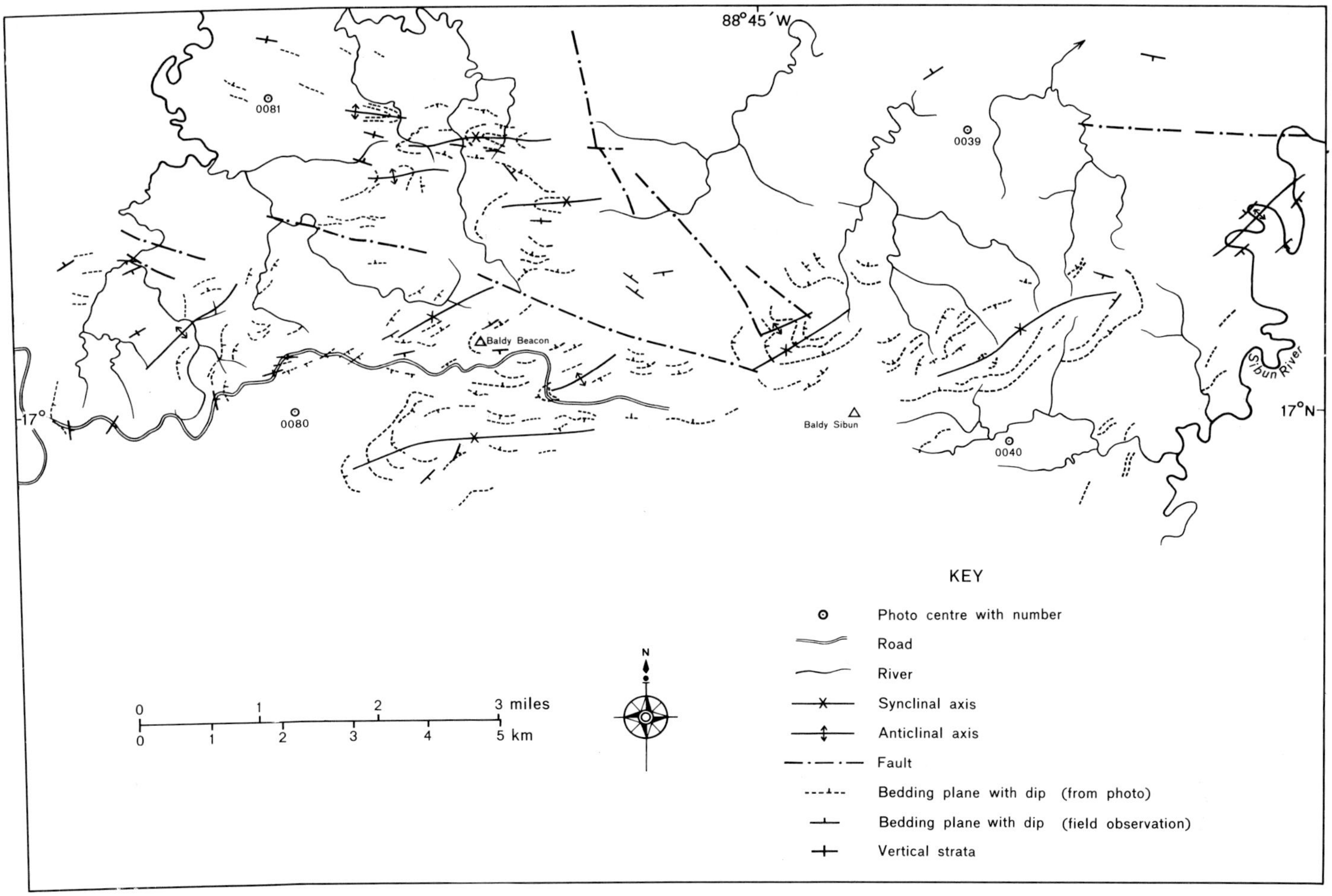

Figure 7 Structure of Santa Rosa sediments in Baldy Beacon area

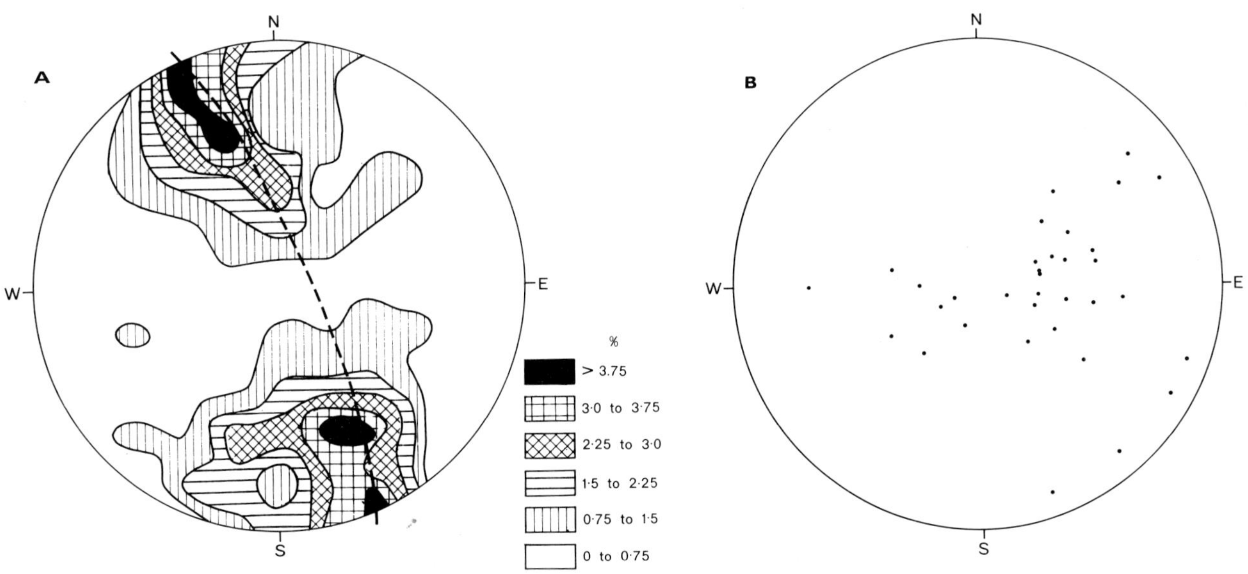

Figure 8
A Contoured projection of 324 poles to bedding for Santa Rosa Group.
B Projection of 34 poles to fold plunges of later folds

the thin covering of the Santa Rosa Group. In some areas the metasediments are affected by more than one phase of deformation shown either by the refolding of cleavage or by earlier folds in which a later cleavage may be developed. These areas are all in close proximity to either granite intrusions or possible unexposed granite lying close to the surface, or are near known faults. Figure 8B shows a stereoplot of the plunges of minor folds which can be shown to be later than the main phase of deformation. Although the readings are insufficient to be of great statistical value, the plots fall into groupings having axes making only small angles with the axis of the main phase of folding. They may represent later phases of tectonism produced under stress directed similarly to that which caused the main deformation. On the other hand the lack of superimposed structures away from the granites would strongly suggest that they were caused by forceful intrusions of the granites. This is supported by the fact that a moderately well-developed slaty cleavage with strike parallel to the axial planes of the folds, is ubiquitous in the argillaceous rocks, whereas later cleavage is developed only within close proximity to the granite. The uniform scatter shown in Figure 8B may well be due to geographical bias since the eastern and western margins of the granites are either not well exposed or not easily accessible. Consequently nearly all the relevant observations were made around the northern and southern margins of the granites.

FAULTING

The present horst structure of the Maya Mountain Block is believed to have been initiated in Middle Cretaceous times, although it is generally agreed (McBirney and Bass, 1969, p. 270) that major regional lineaments were inherited from features of at least Palaeozoic age. Since Middle Cretaceous times the block has been subjected to periodic relative uplift until the mid-Tertiary (Figure 9). Movement along the

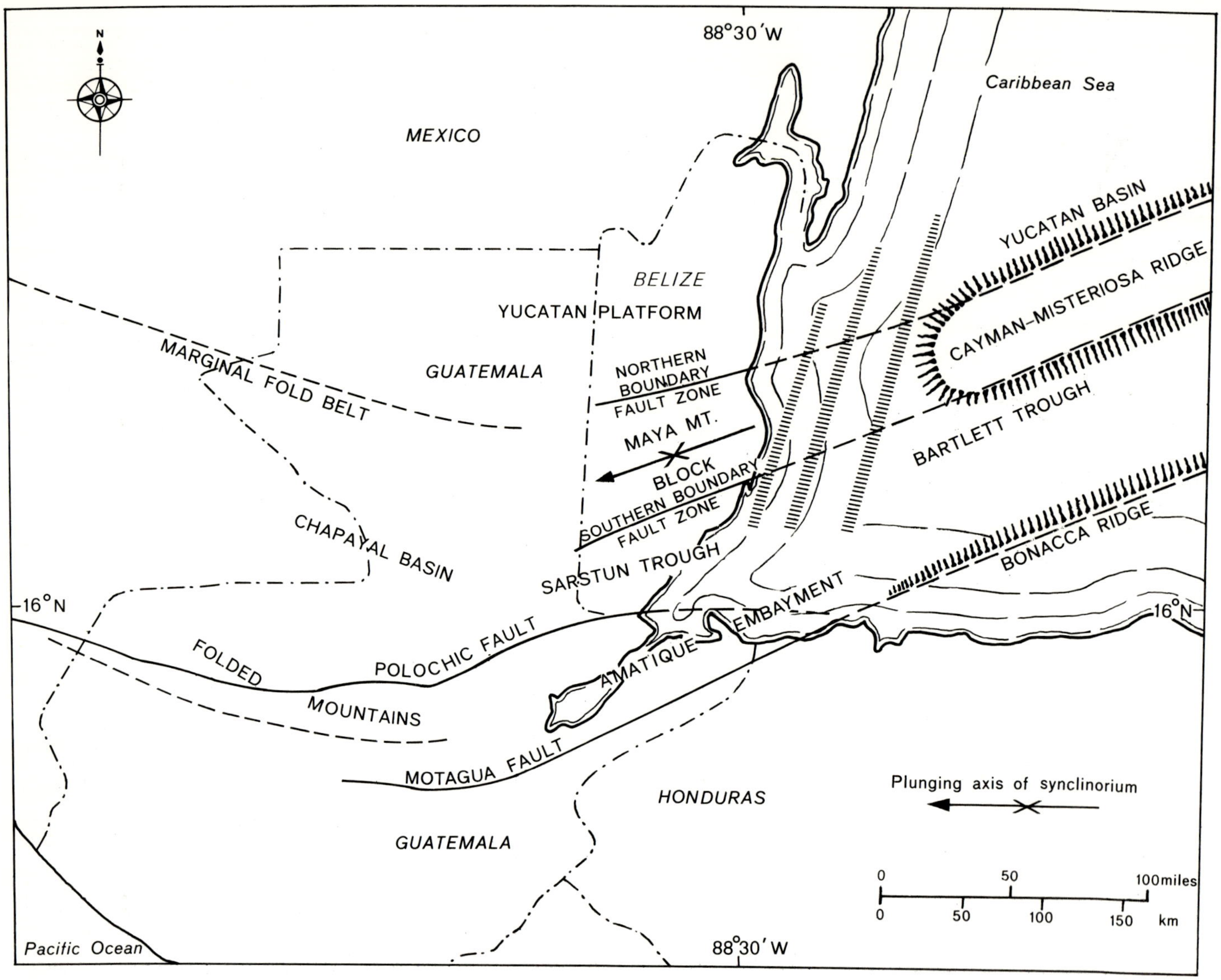

Figure 9 Regional structure of north Central America

northern boundary fault started in Jurassic times allowing the Yucatan Platform to the north to accommodate marine Jurassic sediments which have been recorded in wells (Vinson and Brineman, 1963, p. 104). On the other hand no pre-Cretaceous Mesozoic strata have yet been identified in the Sarstun Trough to the south (Figure 10). The horst is envisaged as having initially had a considerably greater elevation on the east, where only minor overstepping of the boundary faults occurred, in comparison with the west where the metasediments of the Santa Rosa Group and the granite are unconformably overlain by Cretaceous limestone thickening progressively westward. In detail both Northern and Southern Boundary 'faults' are zones of several faults rather than discrete fault planes.

The presence of Eocene and Oligocene sediments off the coast of Belize (Dengo and Bohnenberger, 1969, p. 213), suggests that the eastern boundary of the Maya Block is also controlled by a fault which was active intermittently during most of the Tertiary. Regional faults trending north-north-east parallel to the margin of the mountains along the Inner Channel about 50 km from the coastline have been suggested by Moody in an unpublished report in 1964. A pronounced submarine scarp some 80 km from the coast (Banks and Richards, 1969) may also indicate the position of another major fault zone.

Part of the present-day southern scarp of the Maya Mountains is thought to be an old coastline. It appears that this original fault scarp has retreated northward leaving part of the Southern Boundary fault zone obscured by coastal deposits. Similarly the eastern scarp may well have receded some distance to the west from its original position.

Figure 10
Diagrammatic representation of the development of the Northern and Southern Boundary Faults

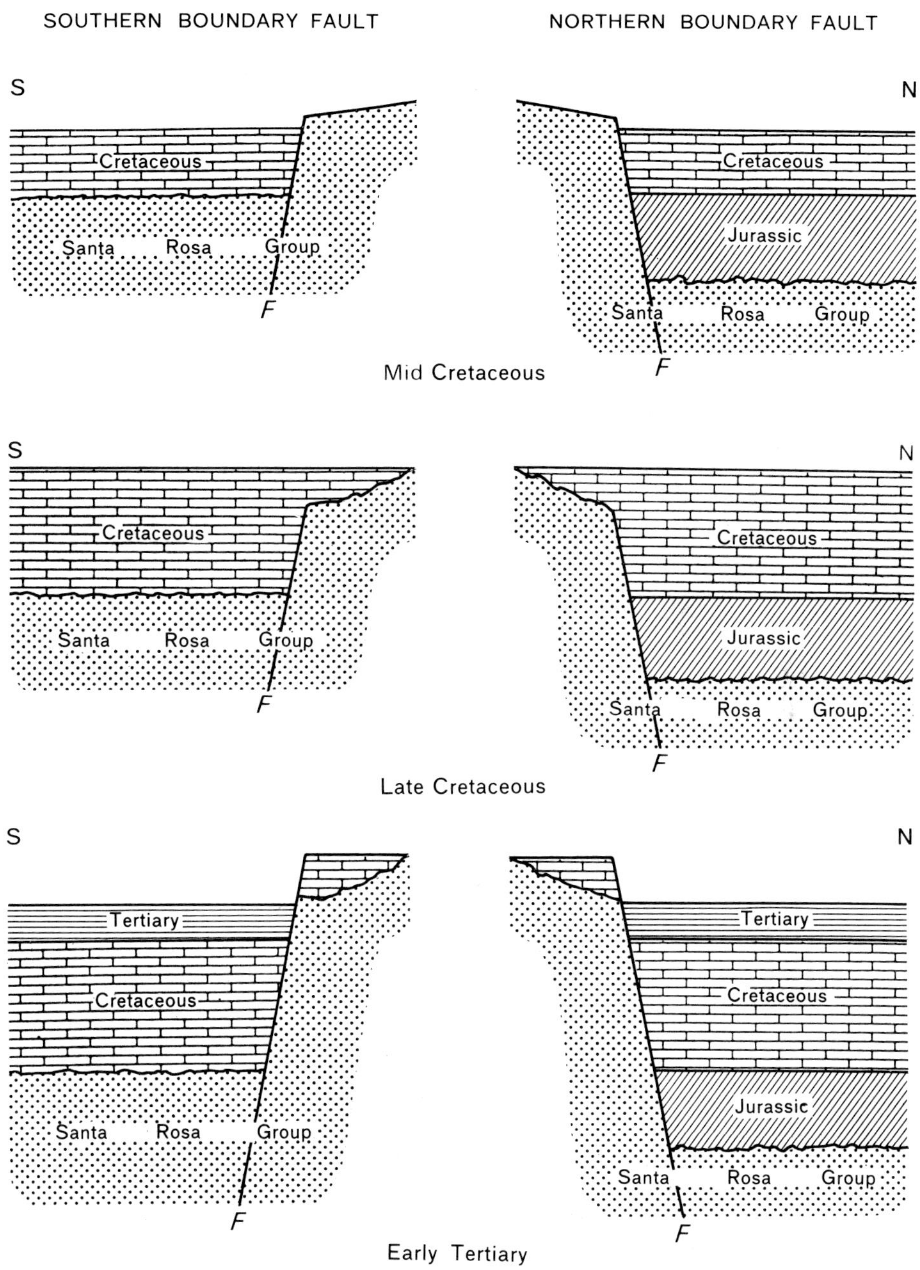

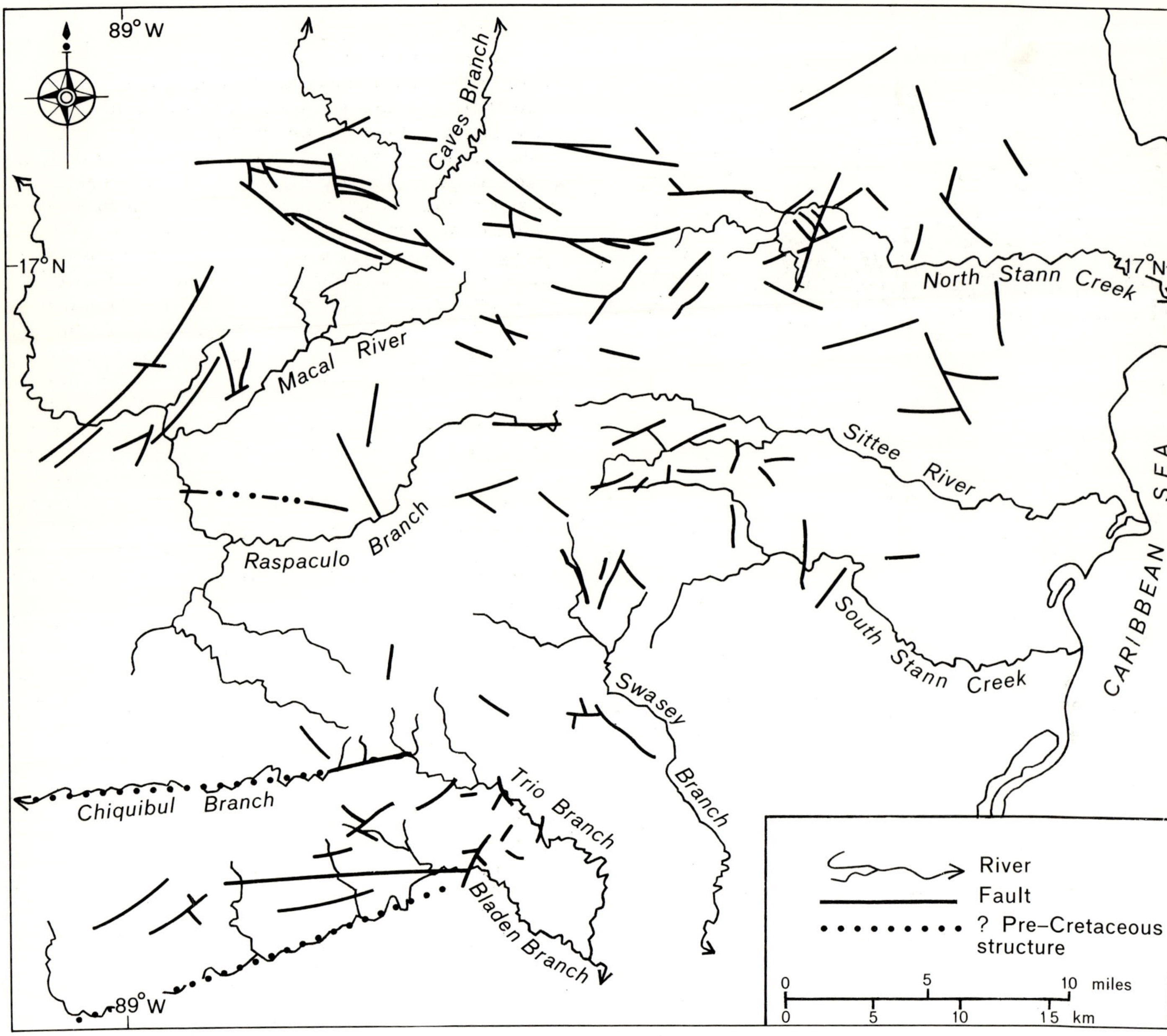

Figure 11 Major faults and fractures in the Maya Mountains

Apart from the boundary faults, the rocks of the Maya Mountain Block are cut by a number of other faults (Figure 11) and have also been considerably affected by jointing. Figure 12 shows azimuth frequencies for (a) joints measured in the field in the Santa Rosa Group; (b) joints in the granites; and (c) faults taken mainly from aerial photographs. Several main trends can be seen in each group of readings. Joints in the Santa Rosa Group show a predominant trend of N135° to N155°, and a less well marked trend of N015° to N035°. Joints in the granites show the same trends but the N015° to N035° trend is predominant. The main trends of faults are N085° to N115° and N155° with less well marked trends N045° to N065° and N175° to N185°. There are insufficient data to relate any of these directions to idealised stress fields, but it may be significant that the dominant fault trend, N085° to N115°, which includes parts of the Northern and Southern Boundary fault zones and most of the largest faults, is the least well developed joint direction and may suggest deeper-seated controls. Some faults have been offset by later faults, but there appears to be no consistent relationship between age and direction. One east–west fault, just north of Chapayal, can be seen to affect the Santa Rosa Group but apparently not the overlying Cretaceous limestone. Some rivers crossing the limestone, notably the Bladen and Chiquibul, follow straight courses for considerable distances. Though no evidence for faulting is seen in the limestone along these rivers, it is thought that they may reflect faulting of pre-limestone age in the underlying metasediments.

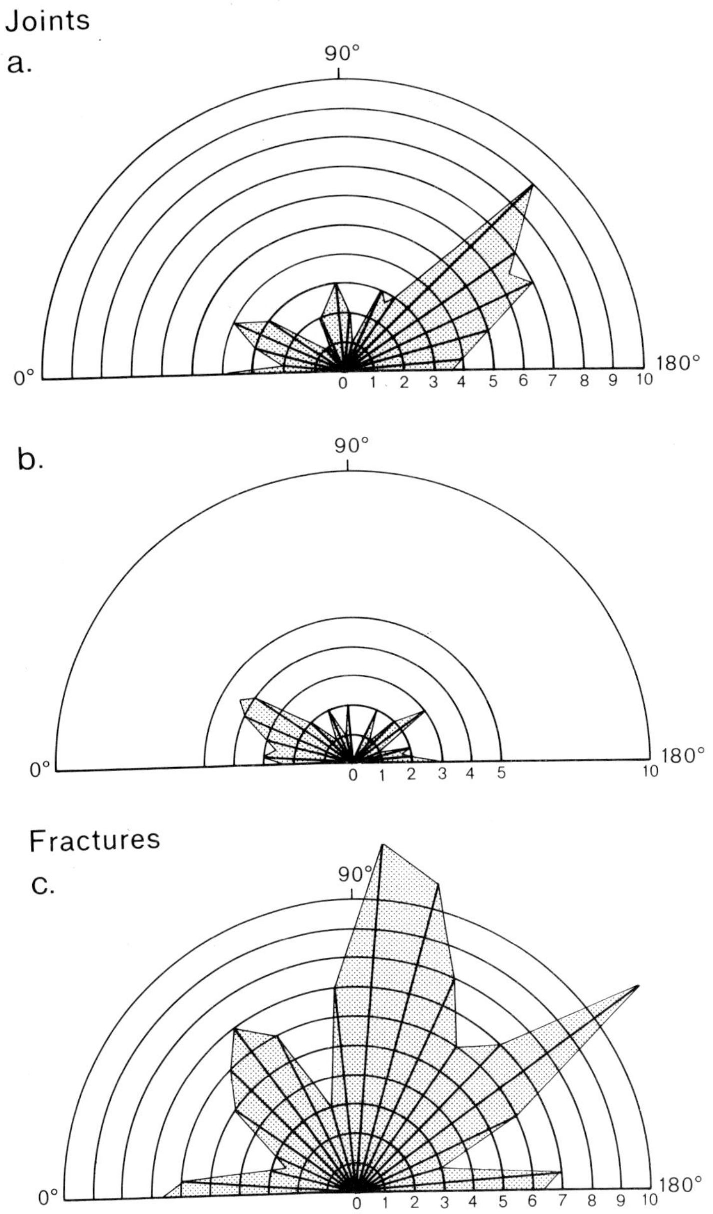

Figure 12 Frequency diagrams for azimuths of joints and fractures.
a. 60 observations, Santa Rosa Group.
b. 35 observations, granites.
c. Fracture azimuth frequency diagram for 118 fractures, taken mainly from aerial photographs of the area.

STRUCTURAL DEVELOPMENT

The oldest and main orogeny which deformed the Maya Mountain Block took place in late Permian and early Triassic times. The youngest rocks to be affected, the uppermost Santa Rosa Group, are palaeontologically placed in the Middle Permian (Dixon, 1956, p. 81) and the oldest unaffected rocks are granitic intrusions which have been radiometrically dated as Triassic with reasonable confidence (Dengo and Bohnenberger, 1969, p. 210). This is consistent with the generally accepted view (Dengo, 1969, p. 315), that an orogeny (equivalent to the Appalachian) strongly deformed the Palaeozoic rocks of northern Central America in the late Permian. The onset of this orogeny is seen in early Permian times in the northern part of the area, when a palaeogeographic change is indicated by sediments of the Maya Group becoming considerably coarser with intraformational conglomerates and minor unconformities.

No sediments of Triassic age are known from Belize, but during this period typical post-orogenic granites were emplaced. Jurassic rocks occur at depth (Vinson and Brineman, 1963, p. 104) on the Yucatan Platform under a thick cover of Cretaceous and Tertiary deposits but sedimentation does not appear to have resumed in the Sarstun Trough until late Cretaceous times. There is no evidence for the deposition of any post-Middle Permian sediments on the main Maya Mountain Block and it would appear that the northern and southern boundary faults were initiated during the Jurassic and Middle Cretaceous respectively and remained active at least until the end of the Eocene. The effects of the Laramide orogeny have not been distinguished in the Maya Mountain Block but movements on the northern and southern boundary faults are presumably related to it.

The Cayman–Misteriosa Ridge (Figure 9) is usually assumed to be the eastward extension of the Maya Mountain Block into the Caribbean Sea (Vinson and Brineman, 1963, p. 104), and this, in turn, can be projected through the Sierra Maestra in southern Cuba to join up with the island arc of the West Indies. The presence of Eocene and Oligocene sediments off the coast of Belize (Dengo and Bohnenberger, 1969, p. 203), suggests that the Maya Mountain Block is separated from the Cayman–Misteriosa Ridge by north-north-east faulting parallel to the present-day mountain scarp and coastline. The bathymetric study by Banks and Richards (1969) shows a pronounced submarine scarp running north-north-east about 80 km off the present coast which may represent the position of the major displacement. They also found (Banks and Richards, 1969, p. 227) only thin recent sediments near the continental margin of the Bartlett Trough lying on normal oceanic crust (Figure 9) and suggest that the Cayman–Misteriosa Ridge was formed by uplift rather than down-faulting of the Bartlett Trough. This thin layer of continental sediments, presumed to lie on oceanic crust in the Bartlett Trough, is in strong contrast to the considerable thickness of Cretaceous and Eocene sediments overlying Maya Group metasediments found in the presumed westward extension, the Sarstun Trough. This suggests that the eastern boundary fault is an important zone of structural weakness with a complex history. Faults with this trend, cutting Tertiary sediments, also occur in northern Belize (Ower, 1928); Dengo (1969, p. 315) postulates transverse faulting parallel to the eastern edge of the Yucatan Peninsula during the mid-Cretaceous. No evidence for lateral movement was found in the Maya Mountains. Banks and Richards (1969, p. 227) show that it is the southern boundary fault and not the Polochic fault that lies along the westerly projection of the northern scarp of the Bartlett Trough (cf. Dengo, 1969, p. 317).

Although it is generally accepted that the Bartlett (Cayman) Trough is the eastward extension of the Motagua–Polochic fault system there are still conflicting theories regarding the amount and direction of displacement along this zone.

The available field data from parts of Guatemala indicate, according to Kesler (1971), that lateral movement is only of the order of 150 km. This is a great deal less than the figure of 800 km postulated as the pre-Miocene movement by Malfait and Dinkelman (1972, p. 259). Pinet likewise argues (1971, p. 2031) that since offsets in excess of 150 km are inferred '. . . from magnetic, topographic and geologic data', along the second-order fault systems (NW–SE) associated with the Bartlett Trough, therefore '. . . much larger dislocations have occurred within the Cayman, sinistral, wrench-fault zone'.

Barr (1974, p. 56) sums up the current position—'several major sub-parallel fault systems occur along this structural axis in Guatemala—the Polochic–Motagua faults—along which lateral movement is suspected but is as yet unproven (Dengo and Bohnenberger, 1969; Kesler, 1971).'

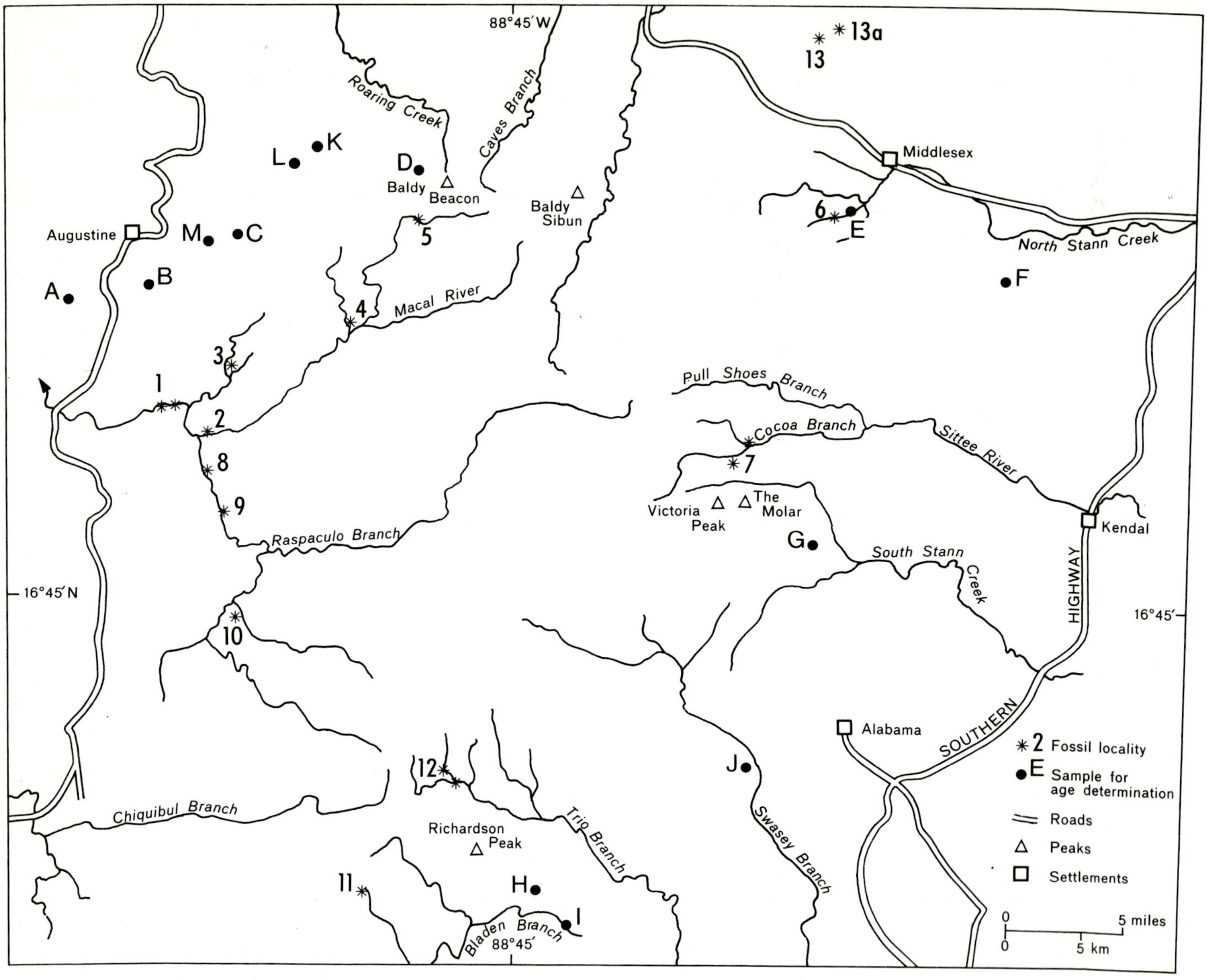

Figure 13 Fossil localities and localities of samples for age determination

SEVEN

Palaeontology

Until Dixon commenced his survey work in the Maya Mountains very little reliable palaeontological evidence had been available. The earlier writers had usually assigned geological age to the sediments on the basis of comparison with rocks of similar lithological type in the surrounding territories, in particular Guatemala.

The Santa Rosa Group sediments have yielded over the years a sufficiently diagnostic collection of fossils for the age range of these rocks to be reasonably well defined (Plate 4, and Figure 13). Dixon found several fossil localities in the black shales and calcareous shales outcropping to the north of the Macal River. From these were identified sufficient genera to date this part of the succession as probably Upper Pennsylvanian. The brachiopods, *Derbyia* cf. *kansasensis*, *Pulchratia* cf. *symmetrica*, *Linoproductus prattenianus*, and *Spirifer opimus*, were considered by Dr Muir-Wood to be of Upper Pennsylvanian rather than Permian age (Dixon, 1956, p. 81). This conclusion is supported by the investigations of S. Ware (Dixon, 1956, p. 81) who has determined various bivalves among which are *Acanthopecten carboniferus*, *Parallelodon* cf. *tenuistriatus*, *Permophorus* cf. *tropidophorus* and *Myalina sp.* Of slightly younger (Lower Permian) affinities were also *Schizodus wheeleri* and *Edmondia*? cf. *terminale*. Identifiable plants are not common, but impressions of *Cordaites*, *Neuropteris* and *Calamites* are also recorded from this same general area (Dixon, 1956, p. 80). During the course of the recently completed survey another fossiliferous locality within the area described above was found but it yielded only indeterminate fragments of bivalves. A new locality in a small stream section about 2.5 km south-west of Baldy Beacon has yielded, from a mudstone unit, some poorly preserved crinoid columnals and fenestellid bryozoa of a type which ranges from Ordovician to Permian.

Slightly better preserved material has also been found in the headwaters of a tributary of the South Branch Creek (16° 59′ N, 18° 31′ W) in a succession which, in addition to the fossiliferous mudstone, contains well-exposed conglomerates and pebbly sandstones. The Department of Palaeontology of the British Museum (Natural History) reports that this material contains crinoid ossicles (not earlier than Ordovician) and indeterminate fenestellid bryozoa (Ordovician–Permian). Possible pterinopectinids, which occur from Viséan–Namurian to Lower Permian and are abundant in the Pennsylvanian, were also identified. The bulk of this data, although imprecise, seems to indicate that a general age of Pennsylvanian can be assigned to these beds.

Further to the south, in the Cocoa Branch of the Sittee River (16° 51′ N, 88° 33′ W), the geological succession is unusual in that it includes some 200 to 300 metres of bedded, sporadically fossiliferous limestone. Specimens of goniatites and foraminifera have been examined but, because of the recrystallisation of the fossils, determination is difficult and non-specific. The goniatites are reported to have 'traces of much distorted suture lines of Carboniferous or Permian type' (Dr M. K. Howarth, written communication, 1970). The fusulinid foraminifera also are unfortunately poorly preserved but it is thought that all the examples seen are representative of a single species. Dr C. G. Adams (written communication, 1970) inclines to the view that the structure of these forms, particularly the spirotheca, is sufficiently characteristic for them to be identified as belonging to the subfamily Schwagerininae and he would refer them to the genus *Parafusulina*—a well-established Permian form. The identification of these foraminifera as Permian is important since this rock unit can now with reasonable confidence be equated to the Chochal Formation of parts of Guatemala which has, as part of its characteristic fossil assemblage, a number of species of the genus *Parafusulina* (Hoffstetter, 1960, pp. 56–58).

A number of samples of poorly fossiliferous mudstone from the upper parts of the Trio Branch (16° 39′ N, 88° 46′W) and the Quebrada de Oro creek (16° 35′ N, 88° 51′ W) have yielded badly preserved fusulinid foraminifera and crinoid columnals which, although the genera are indeterminate, are considered to be most likely of Upper Palaeozoic age (R. P. S. Jefferies, written communication, 1970). One of the conglomeratic limestones from the Trio Branch area contained a lonsdaleiid coral, specimens of *Schwagerina sp.* and a single rather poor specimen of ?*Verbeekina sp.* which are regarded as probably of Permian age (C. G. Adams, written communication, 1970). As a result of these determinations the outcrops in the area immediately to the north of the Bladen Volcanic Member can now be regarded as being probably no older than Upper Pennsylvanian and very likely to be mainly Permian in age.

It is of considerable interest here to note the results of an investigation conducted by C. A. Ross in 1962 on a collection of fossiliferous material obtained by L. H. Ower in 1928 from a locality in the Trio Branch. He identified *Schwagerina gruperaensis*, *Schwagerina sp.* A and *Eoverbeekina* aff. *E. americana* together with other forms. He concludes that this assemblage is closely similar to that reported by Thompson and Miller (1944) from the Grupera Formation of southern Chiapas, Mexico (Ross, 1962, p. 299), and that the species *S. gruperaensis* is well known from the Chochal Limestone of Guatemala (Ross, 1962, p. 299).

The identification of the cephalopod *Perrinites hilli* (orginally by Chubb and confirmed by Spath) obtained by Dixon (1956, p. 20) from a boulder of sandstone in the Raspaculo Creek, about 4 km from its junction with the Macal River, was the first indication that some of the rocks in the Maya Mountains are of Middle Permian age. This has been confirmed during the recent investigation when another specimen of *Perrinites hilli* was found in a tributary of the Monkey Tail Branch about 12.5 km due south of the Dixon locality. This sample was also taken from a loose block but it could not have travelled far; similar specimens in a poor state of preservation were observed in situ. The discovery of this fossil (which was accompanied by unidentifiable bivalves and bellerophontoid gastropods) in this locality indicates that sediments of the Middle Permian extend much further south than Dixon himself envisaged (Dixon, 1956, pp. 20 and 81).

No new palaeontological data are available for the limestones of the Cretaceous. It appears quite well established that the data obtained by Flores (1952) to the north of the Maya Mountains is sufficient to support at least an Upper Cretaceous age for the massive limestone–dolomite sequence and the underlying arenaceous beds. The few fossils collected by Dixon (1956, p. 81) from the Toledo District, to the south of the area discussed in this report, appear to be uppermost Cretaceous with possibly some Palaeocene. Dr Bronnimann and R. J. Macaulay identified the following: *Cosinella sp.*, *Sulcoperculina sp.*, *Oligostegina sp.*, *Dicyclina sp.*, *Vaughanina cubensis*, *Orbitoides sp.*, *Orbitoyclina sp.*, together with miliolids, mollusc fragments and algae.

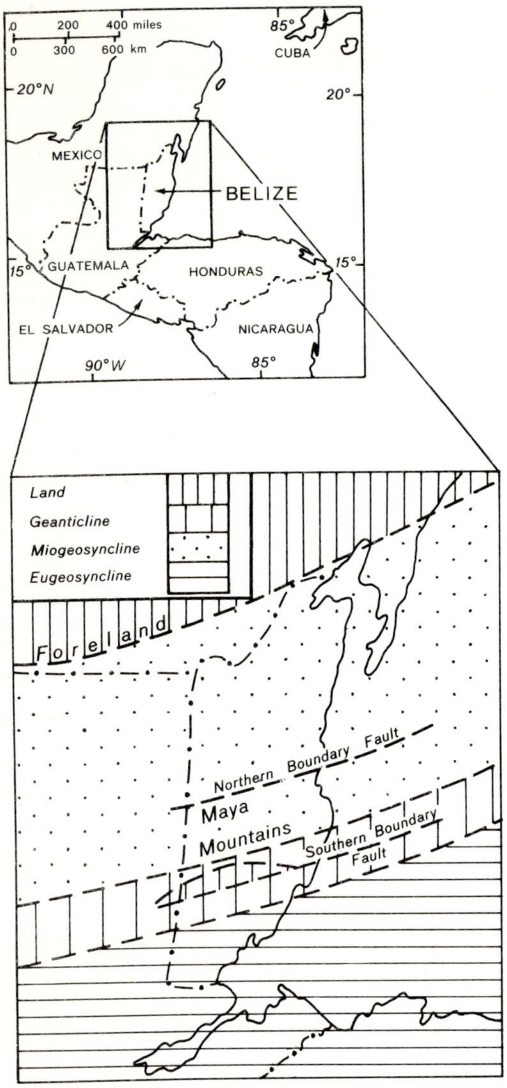

Figure 14 Palaeogeography of Belize and adjacent areas during the Upper Palaeozoic (after Hall and Bateson, 1972)

EIGHT

Palaeogeography

PRE-MESOZOIC

The oldest known rocks in the Maya Block occur along the northern and southern margins and are thought to be contemporaneous. Along the northern margin, in the Baldy Beacon–Baldy Sibun area (17° 00′ N, 88° 45′ W), and immediately to the south of the Mountain Pine Granite (16° 53′ N, 88° 56′ W), the sediments are dominantly arenaceous, suggesting that a coastline running roughly east–west existed at no great distance to the north. Along the southern margin the sediments flanking the Bladen Volcanic Member, particularly in the Trio Branch (16° 38′ N, 88° 47′ W), are arenaceous and are directly related to the northern coastline of a long narrow island formed contemporaneously by the volcanics. Arenaceous sediments also occur in the east of the Maya Block in the headwaters of the South Branch of North Stann Creek (17° 00′ N, 88° 33′ W) and the Blackwater Branch (16° 55′ N, 88° 32′ W), suggesting that a coastline existed somewhere to the east. Thus the broad framework controlling sedimentation during the Upper Carboniferous to Middle Permian (Figure 3) appears to have been an elongate basin trending roughly east–west, with coastlines to the north and east, partially cut off from contemporary sedimentation to the south by the Bladen Volcanic island, but freely connected to the west with the main area of sedimentation. The arenaceous rocks in the northern part of the area are dominantly subgreywackes and orthoquartzites, probably derived from a hinterland of some elevation and deposited rapidly with little sorting. Evidence for tectonism in the presumed hinterland to the north is given by the thick polymict conglomerates that occur slightly higher up the sequence in the same area, reflecting a sudden influx of much coarser material.

Along the margins of the Bladen Volcanic island in the south, the lavas are overlain by agglomerates, conglomerates and coarse sandstones composed entirely of volcanic

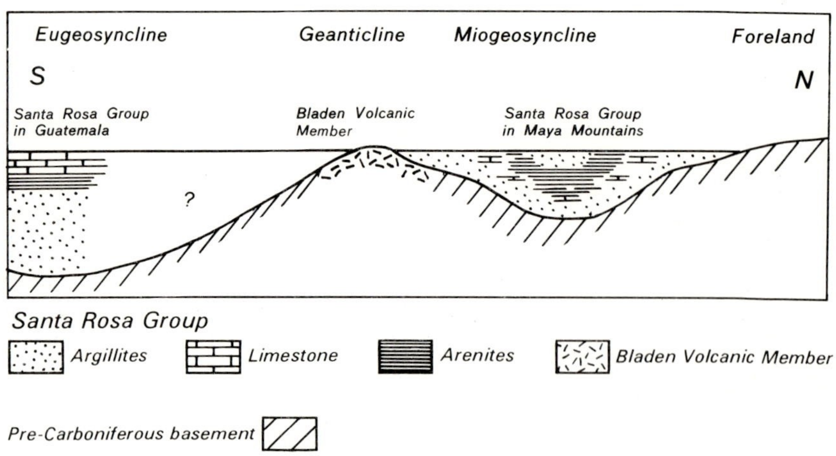

material. These vary from unstratified rocks with completely unsorted angular fragments to fairly well-stratified deposits with some sorting and well-worn pebbles. A gradation from the former to the latter is sometimes observed away from the lavas. These volcanic sediments are overlain by normal sediments and in some places are interbedded with them. The wide variation in sedimentary type and their relationship with the volcanic sediments reflect the diversity of conditions along the coast of the island and also the instability of the area during this period. In some places the volcanic sediments are overlain, commonly with unconformity, by sandstone–seat earth–carbonaceous shale sequences, or by conglomerate–sandstone–argillite sequences. In other places the dominant marginal rock is argillaceous with bands of volcanic sediment interbedded; in these sequences the beds almost always show slump structures. A small subsidiary volcanic island has been identified to the north of the main island which appears to have commenced with submarine vulcanism and which was gradually built up above sea level for at least long enough to allow sandstone–seat earth–carbonaceous shale sequences to develop around its margins.

Limestones were laid down in several places offshore from the marginal arenaceous rocks. They are all dominantly crinoidal and, with the exception of the much thicker sequence in the headwaters of the Cocoa Branch (16° 51′ N, 88° 33′ W), are impersistent and thin. Many of them show evidence of being clastic deposits. Conglomerates of limestone pebbles indicate that stable conditions were short-lived. The limestones are usually interbedded in the argillaceous sequence which continues across the rest of the main sedimentary basin of the Maya Mountains. The complex structure of the Maya Block makes facies changes difficult to establish in a north–south direction, but Dixon (1956, p. 18) demonstrated that rocks of the same horizon became more arenaceous when traced eastwards along strike in the Rubber Camp area (16° 51′ N, 89° 01′ W). Dark carbonaceous mudstones, in which finely comminuted plant debris can locally be identified, are interbedded with the normal argillites in some parts of the basin. These carbonaceous mudstones are found at various levels throughout this sequence indicating that subaerial conditions existed at no great distance from the point of deposition. In the south the carbonaceous mudstones are rich in sedimentary pyrite and suggest that euxinic conditions were established locally.

Arenaceous sediments are developed around the junction of the Raspaculo and Monkey Tail Branches (16° 46′ N, 88° 54′ W) which lies in the axial part of the region. These sediments are interpreted as being the shallow-water sediments representative of the last stage of a sedimentary cycle during which the basin of deposition was being infilled.

Wells sunk on the Yucatan have proved the presence in the south of Permo-Carboniferous sediments at depth and their absence in the north (Dengo, 1969, p. 314). This led Dengo and Bohnenberger (1969) to postulate the existence in this area of a roughly east–west shoreline during part of the Palaeozoic. This is compatible with most of the evidence from the Maya Block although the occurrence of polymict conglomerates suggests that the shoreline was not as far north during their deposition.

REGIONAL SIGNIFICANCE

The pre-Mesozoic palaeogeography of northern Central America is difficult to establish because of the isolated nature of the outcrops and the lack of detailed correlation between the separate areas. Dengo and Bohnenberger (1969) suggest that an orthogeosyncline, trending east–west, was sited in this area during the early Palaeozoic. McBirney and Bass (1969) suggest that the metamorphic rocks of the Chaucus Series and the El Tambor Formation which occur in Central Guatemala are early Palaeozoic, and eugeosynclinal in origin. These are overlain by members of the Santa Rosa Group which are described as a 'flysch-type series' by Vinson and Brineman (1963) and consequently may be regarded as also being eugeosynclinal in origin. King (1969), on the tectonic map of North America, shows the part of the Santa Rosa Group, formerly referred to as the Macal Series, as miogeosynclinal. The results of the present survey suggest that the whole of the Santa Rosa Group found in Belize was deposited in a miogeosynclinal trough which was bounded to the south by the Bladen Volcanic Member (Hall and Bateson, 1972). The recognition of the Bladen Volcanic Member as an integral part of the succession rather than as a younger intrusion throws additional light on the region as a whole (Figure 14). It occupies the natural position of a geanticline separating the miogeosynclinal Santa Rosa Group in Belize from eugeosynclinal rocks having the same age found further south in Guatemala. The largely subaerial origin of the Bladen Volcanic Member and its relationships with the surrounding sediments are in keeping with its being a geanticlinal ridge. Consequently it can be suggested that northern Central America was the site of a geosyncline which continued to develop throughout most of the Palaeozoic and which can, at least in the later stages, be divided into a eugeosyncline and a miogeosyncline.

MESOZOIC AND TERTIARY

After the filling up of the Upper Carboniferous–Middle Permian basin there was a long break in sedimentation. There is no evidence for the desposition of any later sediments over the greater part of the Maya Block. During the Upper Permian and Trias orogenic activity took place and was followed by the intrusion of granites. This was followed in Jurassic times by the deposition of clastic rocks on the Yucatan Platform to the north of the Maya Block. There is no evidence for the deposition of these rocks on the Maya Block itself, and it seems probable that the Northern Boundary Fault zone controlled the southern limit of these deposits. The clastic rocks are succeeded on the Yucatan Platform by carbonate and evaporite rocks of Cretaceous age (Figure 15). Carbonate sedimentation became much more widespread during the Upper Cretaceous, and limestones were deposited in the Sarstun Trough to the south, on the western flank of the Maya Block and on the Yucatan Platform to the north. Some near-shore deposits (Flores, 1952) occur in the Caves Branch–Margaret's Creek area (17° 06′ N, 88° 40′ W) underlying the Upper Cretaceous limestone, and probably represent the earliest stages of overlap of the northern boundary fault. During the early stages of limestone deposition the Northern and Southern Boundary Faults appear to have controlled the limits of sedimentation but, as the limestone accumulated, thin deposits overlapped on to the main Maya Block. Along the western flanks there was complete overlap although detailed mapping has shown that the western extension of the Maya Mountains was a topographic high during the Upper Cretaceous (Dengo and Bohnenberger, 1969, p. 212). This limestone thins out eastwards onto the main Maya Block which in the east must have had greater elevation, and appears to have been emergent. This widespread limestone sedimentation indicates deposition in shallow-water shelf seas which were receiving very little clastic material. Consequently the Maya Block could not have been exposed to much physical weathering and was probably not greatly elevated. Carbonate and evaporite sedimentation continued through the Eocene to the north of the Maya Block, but to the south more terrigenous material was deposited. The provenance of the pebbles in the basal conglomerate of the Toledo Beds (Eocene) indicates transportation from the south and may reflect contemporary movement on the Polochic or Motagua faults.

Eocene and Oligocene deposits have been identified off the coast (Dengo and Bohnenberger, 1969, p. 203) and the absence of the latter from present-day land areas suggests that the eastern boundary fault zone was active during this period. A limestone conglomerate occurring between Cayo (outside the area) and Roaring Creek (17° 06′ N, 88° 52′ W) may represent a talus deposit formed during an Oligocene emergence (Flores, 1952).

A variety of post-Eocene deposits occur to the north of the Maya Block but there are insufficient stratigraphical data to interpret their palaeogeographical significance.

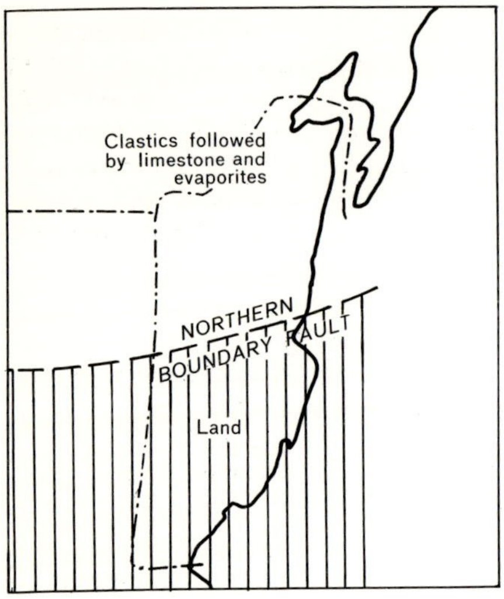

1 Jurassic, Lower and Middle Cretaceous palaeogeography

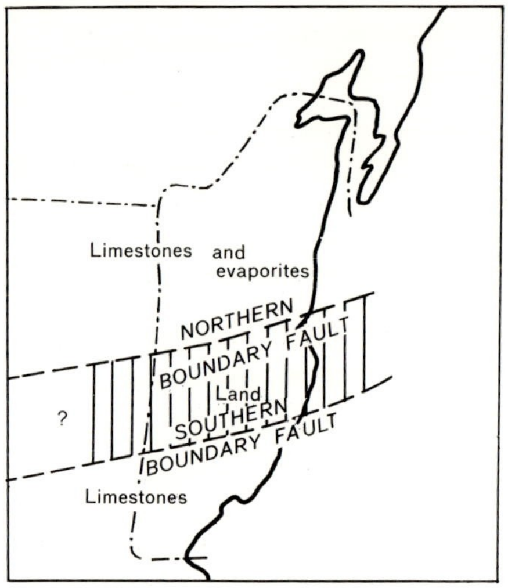

2 Early Late Cretaceous palaeogeography

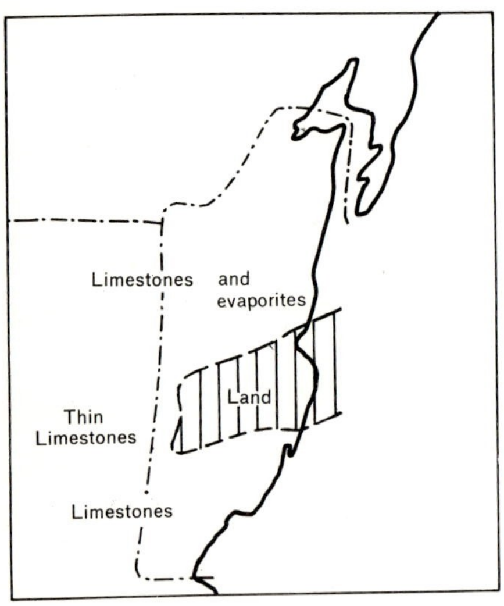

3 Late Late Cretaceous palaeogeography

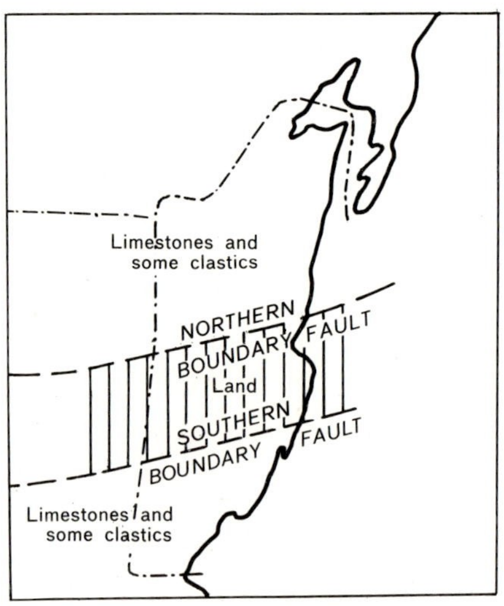

4 Early Tertiary palaeogeography

Figure 15 Post-Palaeozoic palaeogeography of Belize

NINE

Relationship of the geology of Belize to adjacent territories

The relationships of the pre-Cretaceous rocks of the Maya Mountains, the Santa Rosa Group and a number of granitic masses, to rocks of similar age in the rest of the Central American area is in some respects speculative since these rock groups have no continuity of outcrop between the territories. Later Cretaceous and younger sediments have completely buried the older formations over the greater part of the isthmus with the notable exceptions of the Maya Mountains and of the pre-Cretaceous 'basement' rocks exposed between and adjacent to the Polochic and Motagua faults in central Guatemala (Figure 16). In the Yucatan Province of Mexico and the northern part of Belize the pre-Cretaceous rocks are known only from data obtained from oil drilling. In most cases the age of these rocks is based on little other than lithologic characteristics.

Between the Polochic and Motagua faults in central Guatemala the Chaucus Series of metamorphic rocks crops out. These rocks are dominantly garnet-biotite-schists with a structural pattern so complex that the thickness and relationships of these metasediments is unknown. The upper age limit is determined by the fact that the series lies stratigraphically beneath the Santa Rosa Group for which a probable Pennsylvanian age is claimed (McBirney and Bass, 1969). A sequence of mainly metavolcanic rocks, the El Tambor Formation, has been recognised just to the south of the Motagua fault zone. These rocks are thought by some workers to be '. . . broadly equivalent to the higher-grade metamorphic rocks of the Chaucus Series and represents a more volcanic facies of the same eugeosynclinal assemblage.' (McBirney and Bass, 1969, p. 278). The age-determination data, obtained from strontium isotope analyses, do not, however, allow an unequivocal assignment of a Palaeozoic or a Mesozoic age (McBirney and Bass, 1969, p. 274), and thus, in the absence of definitive field data, the age of this formation must be regarded as tentative.

The Santa Rosa Group, '. . . generally believed to be Pennsylvanian or early Permian in age' (McBirney, 1963), contains in its basal members appreciable detritus from granitic and metamorphic terrains. McBirney proposed, on lithological grounds, that the Santa Rosa Formation be subdivided into: a lower series of sandstones and conglomerates, retaining the name Santa Rosa Formation; and an upper division of shales, phyllites and some dark-coloured

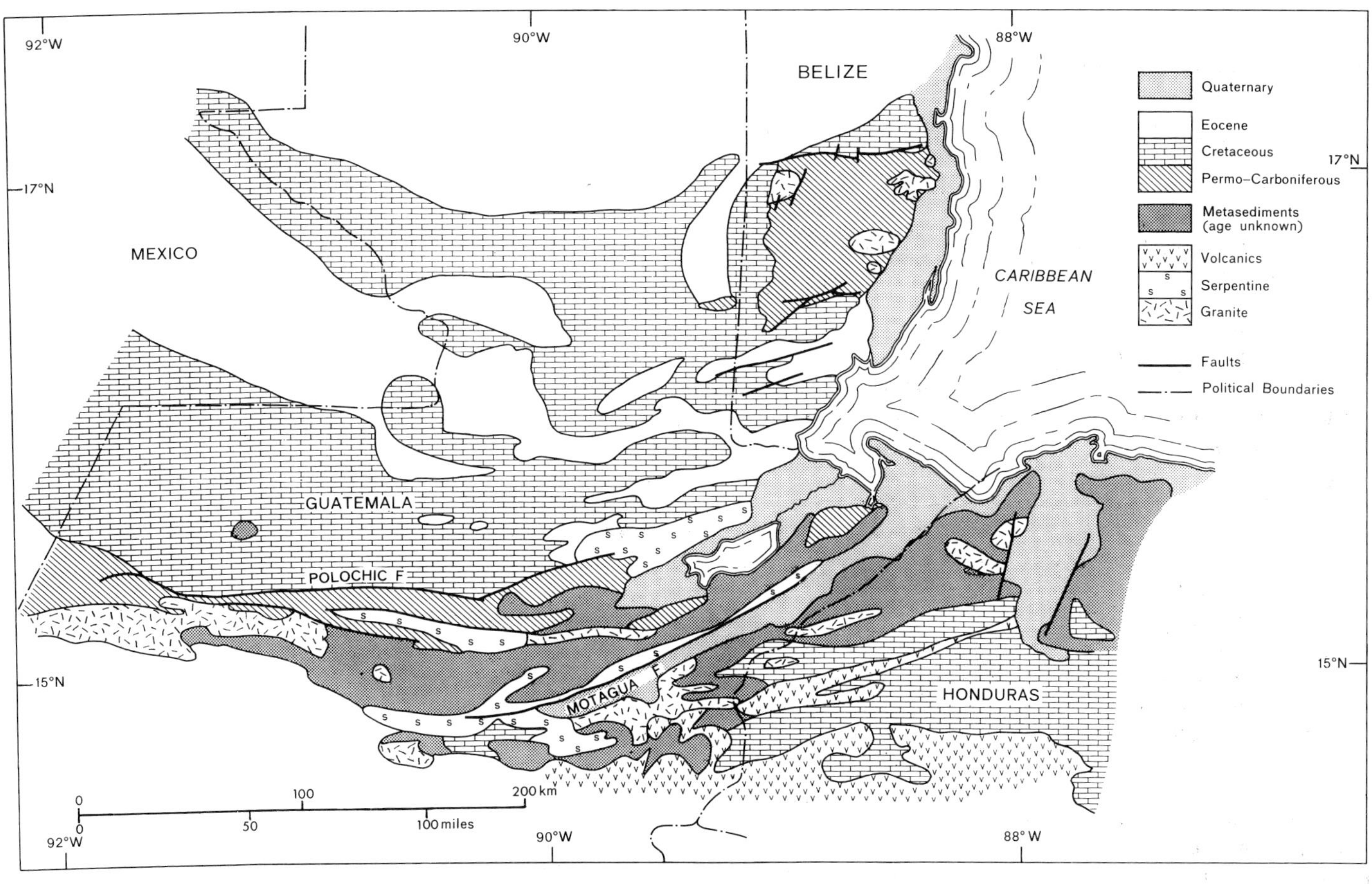

Figure 16 Regional geology of north Central America

limestones, called the Tactic Formation. The name of the latter formation had been coined and used by Walper (1960) in describing a dominantly shaly sequence overlying the more typical arenaceous Santa Rosa sediments in the Tactic Valley, Guatemala. The boundary between the two is based on lithological data, since palaeontological evidence is lacking, and is placed as the position in the sedimentary sequence where the arenaceous facies is replaced by more argillaceous material.

Neither the Santa Rosa nor the Tactic Formation have yielded a fauna suitable for reliable age correlation. However, since the latter passes transitionally up into the more massive Chochal Limestone with a well-established Lower Permian (Leonardian) age, the Tactic is considered probably to be Permian and the Santa Rosa of Pennsylvanian age. This dating corresponds also to the ages given to sediments with similar lithologic and stratigraphic characteristics that have yielded identifiable diagnostic faunas in the Chiapas region of Mexico.

The relationship of the Santa Rosa–Tactic Formation to the metamorphosed Chaucus Series is imperfectly understood. Almost invariably the outcrops of each group are separated by a zone of metasediments which are neither typical Chaucus nor Santa Rosa rocks. It appears that nowhere have Santa Rosa rocks been found in contact with definitely proven Chaucus metasediments; accordingly the boundary between the two is placed rather arbitrarily somewhere in the transitional zone. McBirney has 'adopted the expedient of placing rocks containing significant amounts of brown biotite or garnet in the Chaucus Series . . .' (McBirney, 1963, p. 199). In one area it is reported (McBirney, 1963, p. 200) that the Santa Rosa sediments unconformably overlie the Rabinal granite which is regarded (McBirney and Bass, 1969, p. 273) as contemporaneous with the metamorphism of the Chaucus Series, that is, probably Lower Devonian in age.

The correlation of the pre-Mesozoic rocks of Guatemala with those of the southern part of Belize seems now to be fairly positive (Figure 17). The Santa Rosa Group of Belize, which consists of arenites and argillites with significant limestone horizons, appears to be lithologically compatible with the Santa Rosa–Tactic–Chochal succession and is more reliably dated by its fauna. The palaeontological evidence confirms a Pennsylvanian–Middle Permian age for the Santa Rosa Group which corresponds well with that suggested for the Guatemalan rocks.

The most obvious difference in the successions of the two areas is the lack of a major limestone sequence in Belize of comparable importance to the Chochal of Guatemala. However, the cliff-forming limestones of the Upper Sittee River (Cocoa Branch) have yielded some fusulinids which, although too badly preserved to be specifically determined, could well be early Permian in age. These limestones may thus be similar to the Chochal both in lithology and in stratigraphic position. There are no known equivalents of the Chaucus Series metasediments exposed in the Maya Mountains.

The correlation of the acid intrusives of the Maya Mountains with those of neighbouring Guatemala is not so obvious. In Guatemala there are a number of acid intrusives varying in composition from granite to granodiorite and quartz-monzonite; the majority of these masses are found in the central region, the Sierra de los Cuchamantes and the Sierra de las Minas. Various ages have been suggested for these. Roberts and Irving (1957, p. 25) report an unfoliated granite cutting a shale in 'the Santa Rosa Formation which . . . is thus younger than Permian(?).' (Note that Roberts and Irving use the term Santa Rosa Formation in the wider sense, that is, to include McBirney's Tactic Formation of (?)Permian age.) More recent data, however, suggest that this particular granite could just as well be post-Permian. The Rabinal Granite was probably undergoing erosion in pre-Santa Rosa times; McBirney and Bass (1969, p. 273) quote a Lower Devonian age for this granite. The oldest apparent age of 280 Ma is quoted for the Matanzas Granite; this is an age approximately equivalent to that determined for the acid porphyry of the Bladen Volcanic Member, and would place these igneous rocks at the Carboniferous–Permian boundary.

Mineralogically the granitic rocks of Belize are similar to those of Guatemala but recent determinations have produced some ages rather younger than any established for other granites in north Central America.

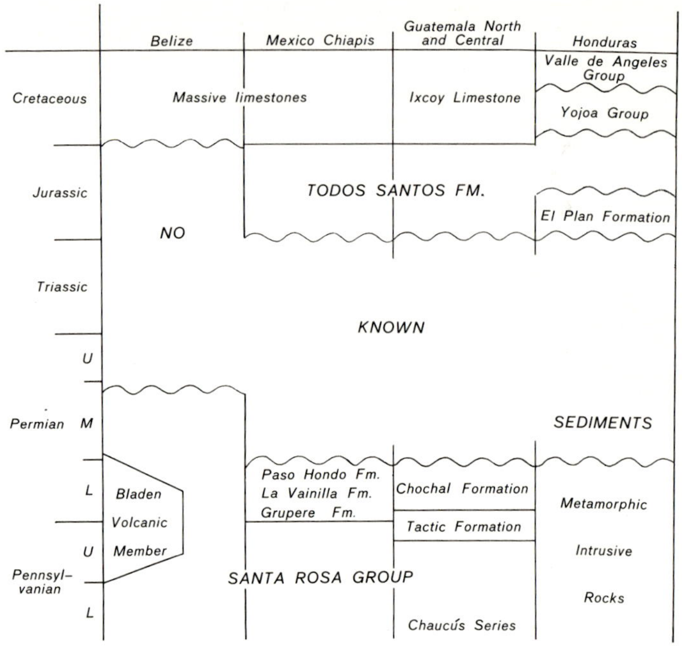

Figure 17 Generalised correlation of the stratigraphy of part of Central America (after Bateson, 1972)

TEN

Economic geology

Until 1950 little had been undertaken in the systematic evaluation of the economic mineral potential within the area of the Maya Mountains. The early workers, such as Sapper and Ower, noted the presence of gold and tin but there was neither sufficient incentive nor local expertise to follow up these slender leads. The secondment of Dixon to the territory had the effect of placing greater emphasis on the economic aspect of the geology, and in the course of his work there he made considerable progress in the further investigation of reported occurrences of minerals of economic significance.

The technique of geochemical prospecting had not, until 1967, been applied to the southern mountainous parts of the territory and by 1971 no applied geophysics, either airborne or ground, had been undertaken elsewhere than in the north and east and in the offshore coastal waters which are regarded as prime targets for oil and gas exploration.

During the period 1967–1969, D. R. Derry Ltd undertook a geochemical prospecting programme in the Maya Mountains under the terms of a prospecting licence issued by the British Honduras Government. Their interest was primarily in gold, copper, lead and zinc. In 1969 and 1970 a team from the Institute of Geological Sciences carried out further geochemical prospecting in conjunction with a geological remapping programme. The stream-sediment samples (Plate 12) collected during this work were analysed for copper, lead, zinc, silver, molybdenum, arsenic, niobium, beryllium, boron and tin.

The detailed results of the analyses of this material are recorded elsewhere (Bateson and Hall, 1970, unpublished). They did not, however, indicate any areas of significantly high values and it was concluded that there was little evidence to suggest any major economic mineralisation in the Maya Mountains.

GEOCHEMICAL PROSPECTING

Although as great a coverage as possible was attempted during the 1969 and 1970 surveys there was a tendency to collect more samples from those areas that seemed, on the basis of the known geological facts, to offer the best conditions for mineralisation. The granites, their aureoles and the mass of the volcanics in the south, were considered to be the most likely areas and consequently were primary targets for sampling. However, sufficient samples were obtained from the rest of the dominantly sedimentary area to provide adequate statistical background data. A simple statistical treatment of the data to produce a value for the mean and standard deviation was undertaken, and for the purposes of this project a value of the mean plus two standard deviations ($\bar{x}+2\sigma$) was selected as the threshold above which the samples were regarded as anomalous.

This treatment of the data has shown that there are only very few samples for which anomalous values for any of the elements are recorded, and many of these are of little significance from a potential economic viewpoint since they are usually geographically isolated. The only potentially interesting results are those for molybdenum which reach values of over 20 ppm in the stream samples from a relatively small and easily defined area in the metasediments forming the aureole of the Cockscomb Granite near to the Cabbage Haul Gap (16° 46′ N, 88° 21′ W). This figure represents a thirteen-fold increase over the average crustal value of 1.5 ppm (Taylor, 1964) and a tenfold increase over the local threshold value. This occurrence of higher molybdenum values was also found to be associated with higher values for arsenic. Insufficient work has been completed in the area to allow the origin of the molybdenum to be stated with certainty but it is tentatively assumed that it is most likely to be associated with the quartz veining that occurs in this part of the aureole. Further investigation into the occurrence of the molybdenum in this area was undertaken by D. R. Derry Ltd, who concluded that the high molybdenum values were the result of the 'scavenging' effect of graphitic shales in the aureole of the granite.

The rest of the geochemical values obtained do not suggest that there is any mineralisation of economic importance, but in some instances they show a distinct relationship to the geology. The outcrop of the Bladen Volcanic Member is marked by a concentration of samples that are anomalous for lead and beryllium. In the case of both of these elements the values accord reasonably well with the figures quoted (Taylor, 1964) for the content in felsic rock types (approximately 48 and 6 ppm respectively). These contrast with the values to be expected in sediments (20 ppm and 3 ppm respectively) and granites (40 ppm and 4.5 ppm). The results of analyses for lead and beryllium in samples from a tributary of the Chiquibul Branch (16° 37′ N, 88° 58′ N) have helped to confirm that there is also a small extension of the Bladen Volcanic Member in that region.

High beryllium values are found in a small area of the Hummingbird Granite in the vicinity of the Mountain Cow Creek (16° 57′ N, 88° 26′ W), where consistently high values delimit a part of the granite as being apparently different in chemical composition from the rest. Surface evidence in this area indicates that there are small pegmatitic dykes and patches here and it is presumed that the excessively high beryllium values are derived from these.

The Mountain Pine Granite is also shown to be different from the other granites in the area in its lack of high boron values, which are common in all the other granites. This is quite clearly related to the tourmaline which is frequently found as an accessory mineral, but which is generally absent from the Mountain Pine Granite except for one small area.

The earlier geochemical prospecting carried out in the area by D. R. Derry Ltd showed that there is an apparent connection between high zinc values and the occurrence of crinoidal limestone in the Santa Rosa Group. Possibly some of the single high values of zinc found during the 1969–1970 project may also be attributed to this source.

Rocks adjacent to both known and supposed faults were quite commonly found to contain pyrite and arsenopyrite.

The geochemical analyses for arsenic show high values, not only associated with molybdenum occurrences, but also with these areas of faulting. In several cases the alignment of the high values suggests an extension of the fault zones beyond the area where they are inferred from the surface geological investigation.

The geochemical data strongly suggest that the Maya Mountain Block is lacking in any major mineralisation and that the majority of the high values found are related to the geochemical characteristics of the lithologies of the various geological units, an exception being the molybdenum at Cabbage Haul Gap (16° 46′ N, 88° 21′ W).

VEIN DEPOSITS

There are several recorded occurrences of economic minerals in quartz veins, a few of which were examined in some detail by Dixon (1956).

Baryte

A zone of quartz veins containing baryte was first located by Dixon in 1951 during the course of geological mapping in the Mountain Pine Ridge area and was examined in greater detail during 1954 and 1955. The occurrence is situated not far from the eastern boundary of the Cretaceous Limestones which here overstep the Mountain Pine Granite, approximately 1 km west of the main all-weather road at San Luis in the south of the Mountain Pine Ridge and on the southern bank of First Creek (16° 48′ N, 89° 02′ W). At the surface the veins are indicated by boulders containing a mixture of quartz, baryte and silicified country rock on a low ridge some 1800 m long and 60 m wide which strikes north-west.

Dixon investigated the deposit by a series of pits and trenches across the zone to obtain information on the grade of the baryte as well as to attempt to identify its origin. He concluded that the baryte mineralisation was a late-stage hydrothermal phenomenon, younger than the granite and the Cretaceous limestone. The baryte is not present in all the quartz veins of this zone. The best veins in the area examined proved to be 1.5 m and 3 m thick respectively, the former on the hanging wall and the latter on the footwall of the zone, which here dips to the south-west at approximately 75°. Channel samples obtained from pits and trenches in the most favourable location were found on analysis to contain between 40 and 50 per cent barium sulphate, but individual hand-picked pieces of the material contain as high as 85 per cent. In other test pits and trenches values for barium sulphate rose only to 12–15 per cent.

In the material obtained from the pits the baryte appears to have postdated the formation of the quartz and was reported by Dixon commonly to have grown around the quartz. The spaces in the vugs and veins are often infilled by pyrolusite, which becomes an important constituent at the top of the water table. In one block of the vein material two small veinlets of apatite were recorded.

As a result of his work Dixon considered that the best material is contained in the two veins, 1.5 m and 3 m wide respectively, that were traced along strike for 250 m to 300 m. He was unable, however, to undertake any investigation that would have given information on their extent in depth, but he considered that this deposit is too small and of too poor a quality to warrant further work; no change has occurred in the economics to alter that opinion.

Copper, Lead, Zinc

Only one well-mineralised quartz vein was found during the course of the fieldwork. This occurs in the Sibun River approximately 10 km to the south-east of Baldy Sibun. Here the shales and mudstones of the Santa Rosa Group strike north-east with fairly uniform steep dips to the south-east and are cut by a 7.5-metre quartz vein which strikes west and dips south and is traceable at the surface for approximately 150 m long strike. The aerial photographs covering the area show the vein to be associated with a fault

Plate 12 Collecting a geochemical sample from stream sediment. The sample is sieved through 30-mesh nylon cloth into a polythene beaker.

zone which can be traced for about 1 km to the west but only for a short distance to the east of the Sibun River before it is itself displaced southwards by a small fault.

The vein is composite, and exhibits several successive phases of intrusion of massive white glassy quartz which for the most part is barren of any other mineral. Joints and planes between successive phases are often coated with manganese and the latter also show well-developed slickensides. A very much later phase of quartz veining is indicated by small barren veinlets cutting through the large vein. The outer zones of the vein are also characterised by a fairly high proportion of included angular fragments of the somewhat silicified country rock. A relatively narrow and somewhat irregular central zone of the quartz vein, up to 45 cm wide, is composed of large crystallised masses of intergrown galena, sphalerite, chalcopyrite and some pyrite. At some places the mineralised core of the vein is bounded by a marginal zone of finely granular galena with minor quantities of the other sulphides. One of the annual reports of D. R. Derry Ltd quotes an analysis of 'a small hand specimen of the mineralisation' containing 10 per cent zinc, 1.28 per cent lead, and 0.17 per cent copper. This is the only mineralised vein found in the area during the recent survey and as such it is unlikely to be of any economic significance.

ALLUVIAL DEPOSITS

Tin

One of the main aims of the work undertaken by L. H. Ower in the period 1926 to 1928 was to investigate the occurrence of potentially economic minerals in Belize. He was concerned principally with cassiterite and investigated its distribution in gravels but did not determine its origin or attempt any assessment of the value of the deposits. Dixon, like Ower, found that cassiterite is a common constituent in almost all the alluvial gravels derived from the granites of the Maya Mountains and he investigated its source and evaluated some of the more promising gravel deposits. He prospected the gravels in the Pomona area (17° 00′ N, 88° 21′ W) of the North Stann valley where a relatively large amount of alluvium is found, derived from a number of exposed bosses of granite and their aureoles. Thirty-five pits were dug in the alluvials of the stream system draining into the North Stann Creek, but the amount of cassiterite recovered in the heavy minerals was rarely more than a trace and no detailed evaluation was undertaken. These results were confirmed by the London Tin Corporation who, in 1952, conducted a small drilling programme in the same area.

The results obtained by both Dixon and the London Tin Corporation were interpreted then as demonstrating that economic quantities of tin are unlikely to exist in the valley gravels. Although the world market price of tin has increased considerably in the past few years there seems no justification for revising this conclusion.

Less extensive investigations were also undertaken by Dixon in the streams draining the Mullins River Granite (17° 03′ N, 88° 25′ W) and in the Cockscomb Basin (16° 45′ N, 88° 38′ W), the latter a rather low area topographically with granitic bedrock. In both areas cassiterite was found to be present in the gravels in only small amounts. As neither of these regions has appreciably alluvium they offer even poorer economic possibilities than the North Stann Creek.

Dixon, again following the findings of Ower, carried out a quantitative investigation in the northern part of the Mountain Pine Ridge close to the boundary between the granite and the adjacent Santa Rosa Group sediments. He obtained the richest concentrates in the Little Vacqueros (17° 03′ N, 88° 58′ W) and Privassion creeks (17° 02′ N, 88° 58′ W). In the vicinity of these streams, he carried out a pitting and trenching programme to investigate not only the alluvium but also the eluvial material and some of the weathered bed rock. During his examination of the area he noted that quartz mineralisation is common, the veins trending north, east and north-east. Most of the cassiterite appears to be associated with those veins striking north-east. Dixon observed that the cassiterite is concentrated as impregnations in the rocks close to the quartz veins. It does not occur in the quartz itself but in the sheared, platy granite on either side of the veins (Dixon, 1956, p. 56).

Two small areas were selected by Dixon for pitting across the granite/Santa Rosa Group boundary in each of which small quartz veins and stringers were noted. In one of these areas trenches dug through the eluvium and topsoil into the decayed bedrock yielded cassiterite. Selective sampling also confirmed that the highest quantities of cassiterite are to be found in the zone parallel to a vein. A measured quantity of decayed bedrock taken from 0.75 m on either side of a 20 cm quartz vein yielded cassiterite equivalent to approximately 4.5 oz per ton (approximately 130 ppm). In this area it was also found that while values for tin can be obtained from anywhere in the granite, cassiterite is only obtained from the Santa Rosa sediments within 150 m of the granite junction and then only associated with quartz veins.

The second area selected was about 5 km to the west on the Little Vacqueros creek and was chosen because quartz veining appears to be better developed than elsewhere along this margin of the granite. Pits spaced at 60-m intervals were dug along north–south lines 180 m apart. The highest results were obtained from material close to those veins which trended approximately N045°. The values obtained from the pit material here were generally lower than the best values obtained from the first area examined. The second area, unlike the first, has extensively developed plateau gravels composed essentially of rounded vein quartz and quartzite pebbles, the latter derived from the Santa Rosa Group. Values obtained from pits dug into this material ranged from 18 to 31 oz of cassiterite per ton (approximately 510–880 ppm) with values at the lower end of the range being the most common. Small patches of stream gravels were also sampled and found to contain up to 8 oz of cassiterite per ton (230 ppm). In neither area was any attempt made to evaluate the amount of gravel available, and Dixon concluded that the gravels even if present in quantity do not appear to be of sufficiently high grade (Dixon, 1956).

Gold

Gold has been known from various parts of the Maya Mountains since scientific observations were first recorded and it is also said locally that the 'ancient Maya Indians' worked gold at some unknown locality in the mountains during the eighteenth and nineteenth centuries.

A number of localities are quoted by Dixon (1956, p. 60), with assay results varying from 1.5 dwt per ton (2.3 ppm) from the Bald Hills, Mountain Pine Ridge (16° 59′ N, 88° 46′ W) to 6 dwt per ton (9.2 ppm) from a boulder in the Mullins River (17° 03′ N, 88° 25′ W). A record of quartz veins assaying 2 dwt per ton (3.1 ppm) at Cattle Landin, Punta Gorda, which lies to the south of the area, is discredited by Dixon (1956, p. 60) because the rocks of that area are no older than Cretaceous and probably much younger.

'Colours' of gold occurring in alluvial concentrates have been recorded from stream systems draining the Mountain Pine Ridge, the Cockscomb Basin and Mullins River area as well as the Ceibo area, to the south-west of Puente Natural, and the Macal–Raspaculo drainage basin (16°55′ N, 88° 45′ W). In all these localities, with the exception of the Ceibo drainage, the gold occurs as rare small flakes. At Ceibo, which lies to the north-west of the area mapped, the river gravels contain much more gold, usually in a coarse and relatively unrounded form but the quantity of gravel is limited and thus it is likely that the area would support only a few individuals using a simple hand-washing technique. Dixon carried out a short programme of pitting here and obtained values which 'averaged a shade less than 2 dwts per cubic yard of gravel . . .' (1956, p. 61). He was unable to trace the gold back to its source but he obtained sufficient evidence to indicate that it was almost certainly associated with the quartz veins containing arsenopyrite which occur in this area close to the volcanic lavas of the Bladen Volcanic Member. An individual boulder of quartz with arsenopyrite from this area yielded the equivalent of 5.5 dwt per ton of gold when crushed.

In the 1968–1969 period the Ceibo area was investigated by D. R. Derry Ltd who confirmed the basic facts as reported by Dixon and also observed that the area had insufficient quantity of suitable auriferous gravel to support any mining enterprise other than a few small hand-washing operations. Our recent work confirmed the occurrence of 'colours' of gold in many of the previously reported areas, none of which appeared sufficiently promising to justify further examination.

HEAVY MINERAL CONCENTRATES

Dixon collected more than 200 heavy-mineral concentrates panned from all areas of the Maya Mountains: the results of the optical examination of these samples are recorded in tabular form in his publication (Dixon, 1956, p. 49, Table 1). During the recently completed survey further heavy mineral concentrates were collected (Plate 13) and examined, the results of which in general conform with the suites of heavy minerals described by Dixon.

The granites yield zircon, ilmenite, magnetite, cassiterite and tourmaline with some monazite. Some rare tiny flakes of gold are recorded associated with the Cockscomb Granite. This is additional to the better-documented occurrences of Ceibo which lies to the north-west. Well-rounded pinkish zircons are almost ubiquitous in any concentrate from the Santa Rosa Group sedimentary rocks which in addition contains ilmenite, and magnetite. Andalusite, chiastolite and epidote, with occasional grains of garnet, are derived from the sediments affected by thermal metamorphism. The Bladen Volcanic Member provides principally ilmenite and zircon.

INDUSTRIAL MINERALS

The economic possibilities of the bulk materials that are to be found in and around the Maya Mountains have yet to

Plate 13 Panning a heavy-mineral concentrate.

be investigated. Dixon in the nineteen-fifties realised the potential of the limestones but had little time to carry out any very detailed survey. He was able to indicate that these rocks, generally regarded as Cretaceous in age, are commonly interbanded with dolomite. The few laboratory studies that he was able to undertake showed that in general the limestones are fairly pure with very little insoluble residue after treatment in hydrochloric acid and are also low in magnesia. The interbedded dolomites, however, were found to contain between 17.96 and 21.78 per cent magnesia (Dixon, 1956, p. 62).

So far there has been only limited use made of either of these materials. Several quarries have been opened into the limestone-dolomite sequence to provide road metal and aggregate for the building of the new capital site of Belmopan. There have been no large-scale surveys to learn the extent of these rocks, and no tests, chemical or physical, to indicate their potential in industry.

Clays

The coastal plain of this southern part of Belize is underlain by much clay and sandy clay. It is suggested that some of these materials may be suitable for bricks and possibly ceramic products. In some areas the clays are distinctly kaolinitic. No investigation of any consequence has been undertaken. A single channel-sample from a pit one metre deep from a locality just west of the Swasey Branch Crossing (16° 32′ N, 88° 33′ W), has been tested. The results indicate that while it has poor colour quality because of iron oxides and contains finely disseminated mica, its plastic index is close to the specification for ball clays and may, if present in sufficient quantity, have a potential use in ceramics (Ref. D. J. Morgan and others, Report No. 52, Mineralogy Unit, Inst. Geol. Sci.). Extensive field investigations and extensive and exhaustive chemical and physical tests would, however, be necessary to assess the economic potential.

REFERENCES

ANON. 1968. Variations in isotopic abundance of strontium, calcium and argon, and related topics. Pp. 79–80 in *16th Annu. Prog. Rep. Dep. Geol. Geophys., Massachusetts Inst. Technol.* U.S. At. Energy Comm. Contract No. AT(30–1)–1381–16.

BANKS, N. G., and RICHARDS, M. L. 1969. Structure and bathymetry of western end of Bartlett Trough, Caribbean Sea. *Mem. Am. Assoc. Pet. Geol.*, No. 11, pp. 221–228.

BARR, K. W. 1974. The Caribbean and plate tectonics—some aspects of the problem. *Verhandl. Naturforsch. Ges. Basel*, Vol. 84, No. 1, pp. 45–67.

BATESON, J. H. 1972. New interpretation of geology of Maya Mountains, British Honduras. *Bull. Am. Assoc. Pet. Geol.*, Vol. 56, pp. 956–963.

— and HALL, I. H. S. 1970. Reconnaissance geochemical and geological investigation of the Maya Mountains of southern British Honduras. *Rep. Overseas Div. Inst. Geol. Sci. London*, No. 16 [Unpubl.]

— — 1971. Revised geological nomenclature for the pre-Cretaceous rocks of British Honduras. *Bull. Am. Assoc. Pet. Geol.*, Vol. 55, No. 3, pp. 529–530.

BLOUNT, D. N. 1969. Stratigraphy and petrology of the Chiantla area, Guatemala. *Publ. Geol. Inst. Centroam. Invest. y Technol. Ind.* (*ICAITI*), No. 2, pp. 67–72.

DENGO, G. 1969. Problems of tectonic relations between Central America and the Caribbean. *Trans. Gulf Coast Assoc. of Geol. Soc.*, Vol. 19, pp. 311–320.

— and BOHNENBERGER, O. H. 1969. Structural development of northern Central America. *Mem. Am. Assoc. Pet. Geol.*, No. 11, pp. 203–220.

DICKERSON, R. E., and WEISBORD, N. E. 1931. Cretaceous limestone in British Honduras. *J. Geol.*, Vol. 39, No. 5, pp. 483–486.

DIXON, C. G. 1956. *Geology of southern British Honduras with notes on adjacent areas.* (Belize: Government Printer.) 85 pp.

DUNLOP, W. R. 1920. *Report on the economic and natural features of British Honduras.* (London: Crown Agents.)

FLORES, G. 1952. Geology of northern British Honduras. *Bull. Am. Assoc. Pet. Geol.*, Vol. 36, No. 2, pp. 404–409.

FOWLER, H. 1879. A narrative of a journey across the unexplored portion of British Honduras. *Rep. Geol. Surv. Belize.* [Unpubl.]

HALL, I. H. S., and BATESON, J. H. 1972. Late Paleozoic lavas in Maya Mountains, British Honduras, and their possible regional significance. *Bull. Am. Assoc. Pet. Geol.*, Vol. 56, No. 5, pp. 950–956.

HARLAND, W. B., SMITH, A. G., and WILCOCK, B. (Editors), 1964. The Phanerozoic time-scale. A symposium dedicated to Professor Arthur Holmes. *Q. J. Geol. Soc. London*, Vol. 120S. 458 pp.

HOFFSTETTER, R. (Editor) 1960. Amerique Latine, Amerique Centrale. *Lexique stratigraphique international*, Vol. 5, Fasc. 2a. 367 pp.

KESLER, S. E. 1971. Nature of ancestral orogenic zone in nuclear Central America. *Bull. Am. Assoc. Pet. Geol.*, Vol. 55, No. 12, pp. 2116–2129.

— BATESON, J. H., JOSEY, W. L., CRAMER, G. H., and SIMMONS, W. A., 1971. Mesoscopic structural homogeneity of Maya Series and Macal Series, Mountain Pine Ridge, British Honduras. *Bull. Am. Assoc. Pet. Geol.*, Vol. 55, No. 1, pp. 97–103.

KING, P. B. 1969. The tectonics of North America. Tectonic map of North America, 1:5 000 000. *Prof. Pap. U.S. Geol. Surv.*, No. 628.

MCBIRNEY, A. R. 1963. Geology of a part of the central Guatemalan Cordillera. *Univ. California Publ. Geol. Sci.*, Vol. 38, pp. 177–242.

— and BASS, M. N. 1969. Structural relations of pre-Mesozoic rocks of northern Central America. *Mem. Am. Assoc. Pet. Geol.*, No. 11, pp. 269–280.

MALFAIT, B. T., and DINKELMAN, M. G. 1972. Circum-Caribbean tectonic and igneous activity and the evolution of the Caribbean Plate. *Bull. Geol. Soc. Am.*, Vol. 83, pp. 251–272.

NORTH, F. K. 1965. The curvature of the Antilles. *Geol. en Mijnbouw*, Vol. 44, pp. 673–686.

OWER, L. H. 1928. Geology of British Honduras. *J. Geol.*, Vol. 36, pp. 494–509.

PETTIJOHN, F. J. 1957. 2nd Ed. *Sedimentary rocks.* (New York: Harper & Bros.) 526 pp.

PINET, P. R. 1971. Structural configuration of the North-Western Caribbean Plate boundary. *Bull. Geol. Soc. Am.*, Vol. 82, pp. 2027–2032.

REED, F. R. C. 1921. *Geology of the British Empire.* (London: Arnold.) 480 pp.

ROBERTS, R. J., and IRVING, E. M. 1957. Mineral deposits of Central America. *Bull. U.S. Geol. Surv.*, No. 1034. 206 pp.

ROSS, C. A. 1962. Permian foraminifera from British Honduras. *Palaeontology*, Vol. 5, pp. 297–306.

SAPPER, C. 1889. Über Gebirgsbau und Boden des nördlichen Mittelamerika. *Petermanns Geogr. Mitt.*, Vol. 27, No. 127, 119 pp.

— 1937. Mittelamerika. In *Handbuch der regionalen Geologie.* Vol. 8. (Heidelberg: Carl Winters Universitaetsbuchhandlung.) 160 pp.

SCHUCHERT, C. 1935. *Historical geology of the Antillean–Caribbean region* (*or lands bordering the Gulf of Mexico and the Caribbean Sea*). (New York: John Wiley and Sons.) 811 pp.

TAYLOR, S. R. 1964. Abundance of chemical elements in the continental crust: a new table. *Geochim. Cosmochim. Acta*, Vol. 28, pp. 1273–1285.

THOMPSON, M. L., and MILLER, A. K. 1944. The Permian of southernmost Mexico and its fusulinid faunas. *J. Paleontol.*, Vol. 18, No. 6, pp. 481–504.

VINSON, G. L., and BRINEMAN, J. H. 1963. Nuclear Central America. Hub of Antillean Transverse Belt. *Mem. Am. Assoc. Pet. Geol.*, No. 2, pp. 101–112.

WALPER, J. L. 1960. Geology of the Coban–Purilha area, Alta Verapaz, Guatemala. *Bull. Am. Assoc. Pet. Geol.*, Vol. 44, pp. 1273–1315.

WILLIS, B. 1912. Index to the stratigraphy of North America. *Prof. Pap. U.S. Geol. Surv.*, No. 71, 894 pp.

APPENDIX 1

Fossil localities

The accompanying map (Figure 13) shows all the recorded locations from which fossiliferous material has been collected and determined. They are numbered on the map and are described below.

LOCALITY 1. The two locations marked on the map and included in this single reference correspond to Dixon's sites F1–F6 (Dixon, 1956, pp. 79 ff.). The succession consists of black shale and local black calcareous shale within a dominantly arenaceous facies of the Santa Rosa Group. Dixon's F6 is an exposure of fossiliferous sandy grey limestone from a short distance south-west of Rubber Camp (16° 51′ N, 89° 01′ W), on the south side of the river. Materials from these sites were examined by the late Drs L. J. Chubb and H. M. Muir-Wood and by Mr S. Ware, who have recorded the following (some generic names have been revised):

Plants *Cordaites sp.* [leaf], *Neuropteris sp.*

Brachiopods *Derbyia* cf. *kansasensis* Dunbar & Condra, *D.* cf. *plattsmouthenis* Dunbar & Condra, *?Eolyttonia sp.*, *Isogramma* cf. *millepunctata* Meek & Worthen, *Linoproductus prattenianus* (Norwood & Pratten), *?Lissochonetes sp.*, *Orbiculoidea sp.*, *Pulchratia* cf. *symmetrica* (McChesney), *?Retaria lasallensis*, *Schizophoria sp.*, *Spirifer opimus* Hall

Bivalves *Acanthopecten carboniferus* (Stevens), *Edmondia?* cf. *terminale* Hall, *Parallelodon* cf. *tenuistriatus* Meek & Worthen, *Permophorus* cf. *tropidophorus*, *Schizodus wheeleri* Swallow

LOCALITY 2. The dominant sediments here are again referred to as black shales with some associated grey crinoidal limestones. Fossils from this locality (Dixon's F10) are much less common than at locality 1 and have been recorded as indeterminate lepidodendroid fragments and '*Spirifer*' cf. *rockymontanus*.

LOCALITY 3. The exposure at this locality is said to be similar to locality 2, a dark grey crinoidal limestone. There are no records of any specific determinations made on the fauna of this locality.

LOCALITY 4. Poorly fossiliferous mudstone with indeterminate casts of bivalves and crinoid columnals.

LOCALITY 5. Lithologically very similar to locality 4, consisting dominantly of a mudstone–shale succession with intercalated sandy units. Dr C. T. Scrutton describes the material as '. . . fenestellid bryozoa of a type which ranges from Ordovician through to the Permian'.

LOCALITY 6. Fossil material was here obtained from a mudstone unit in a sequence containing conglomerates, pebbly sandstones and micaceous sandstones. The mode of preservation is again poor but the remains of plants and bivalves have been identified as *Neuropteris sp.*, *Sanguinolites sp.* and an indeterminate pectinoid. Indeterminate remains of bryozoa and crinoids are also found.

LOCALITY 7. The fossiliferous localities here are within a dominantly allochthonous limestone sequence with occasional interbedded mudstones. It is thought that these beds can be equated with the Chochal Formation of Guatemala. The most useful of the fossils are the foraminifera which, although recrystallised, have been assigned by Dr C. G. Adams to the genus *Parafusulina* and are regarded as indicative of the Permian. Indeterminate goniatite remains with distorted suture lines of Carboniferous or Permian type are also found.

LOCALITY 8. Dixon's locality F11 consists of loose blocks of non-calcareous sandstones found in the bed of the Raspaculo Branch (16° 50′ N, 88° 47′ W). This material, regarded by Dixon as having travelled only a short distance downstream from its source, contained *Perrinites hilli* (Smith), a Middle Permian goniatite.

LOCALITY 9. Black calcareous shale (F12), 5.6 km up Raspaculo Creek. The following identifications were made:

'Chonetes' sp.
?Lissochonetes sp.
?Orbiculoidea sp.

LOCALITY 10. A new locality discovered in 1970 in the Monkey Tail Branch (16° 44′ N, 18° 57′ W). A large rolled and eroded specimen of *Perrinites hilli* (Middle Permian) and poorly preserved bellerophontoids. The dominant lithology is of shaly mudstones.

LOCALITY 11. In the upper reaches of Quebrada de Oro Creek (16° 32′ N, 89° 49′ W) a fossiliferous decalcified mudstone was found in a succession consisting of mudstone–shale–limestone at least 30 m thick. The crinoid columnals and fusilinids are indeterminate but regarded as Upper Carboniferous–Lower Permian in age.

LOCALITY 12. Of the two samples at this locality one has yielded numerous abraded and broken examples of *Schwagerina sp.* and *?Verbeekina sp.*, and a lonsdaleiid coral from an allochthonous limestone. The other material from a decalcified mudstone has yielded an unidentifiable internal mould of a fusilinid and crinoid columnals. All of this material is assigned to the Upper Carboniferous–Lower Permian.

LOCALITIES 13, 13a. The fossil fragments from the first of these (Dixon's F21) were obtained from shale intercalated with conglomerates and were indeterminate. A productoid of unidentifiable genus was found at 13a (Dixon's F20).

APPENDIX 2

Age determinations on igneous rocks

Thirteen rock samples have been analysed and ages calculated by J. E. Rouse and N. J. Snelling of the Isotope Geology Unit, Geochemical Division of the Institute of Geological Sciences. Of these, eleven were obtained from granitic intrusions and two from the lavas of the Bladen Volcanic Member. Figure 13 shows the approximate location of the samples analysed which are listed in Table 5.

The granites which intrude the Maya Mountains Palaeozoic sedimentary and volcanic rocks can be shown in places to be pre-Cretaceous and comprise four main bodies—the Mountain Pine Ridge batholith, a relatively evolved body of granite and potassic quartz-monzonite, and the Hummingbird and Cockscomb batholiths and the Sapote Granite, which consist of less evolved quartz-monzonites and granodiorites (Kesler and others, 1974).

Experimental methods

Rb:Sr ratios were determined by X-ray fluorescence analysis using samples in which Rb and Sr had been determined by isotope dilution analysis as standards. The $^{87}Sr:Sr^{86}$ ratios were measured on a Thomson THN 206 mass spectrometer. Potassium determinations were made using flame photometry with a Li internal standard. Argon was extracted by fusion in a bakable vacuum system and determined by isotope dilution with a ^{38}Ar spike in a GEC/AEI MS10 mass spectrometer. The estimated precisions used in isochron calculations were ± 2 per cent for $^{87}Rb:^{86}Sr$ and ± 1 per cent for $^{87}Sr:^{86}Sr$. Errors quoted with the K:Ar apparent ages take into account uncertainties in the mass spectrometric ratios, the spike volume, the potassium determinations, as well as error magnification associated with the atmospheric argon correction.

Age determinations

The Mountain Pine Ridge Granite. The petrography and chemistry of this body are described by Kesler and others (1974). Six samples of this batholith have been analysed for Rb and Sr and the data are given in Table 3. Also included in this table are the Rb:Sr data for sample JHB 103, collected from the same locality as sample BHK-66 (see locality 66 in Figure 2 of Kesler and others, 1974). According to these authors the porphyry from locality BHK-66 ('D' on Figure 13) is 'considered to be part of' the Mountain Pine Ridge unit and is presumed to be consanguineous and possibly coeval with the granitic rocks of the Mountain Pine batholith.

The analytical data are shown on the isochron plot of Figure 18A. It is not possible to fit a satisfactory isochron to the data points, the scatter of points about the line of best fit being greater than that which can be attributed to experimental error alone. This situation arises when attempts are made to regress samples which are not contemporaneous or consanguineous, or which have suffered open-system behaviour with regard to rubidium and/or strontium.

The petrographic descriptions and geochemical data presented by Kesler and others (1974) indicate that the Mountain Pine Ridge batholith is a relatively homogeneous body. This suggests that the suite of samples is contemporaneous and consanguineosn and leads to the conclusion that the scatter of data points about the best-fit line reflects some degree of chemical alteration which

Table 3 Rb: Sr data, Belize

Sample	Analysed material	$^{87}Rb:^{86}Sr$	$^{87}Sr:^{86}Sr$
JHB 25	whole rock	73.56	1.0611
JHB 25	K-feldspar	104.63	1.1359
JHB 6	whole rock	44.53	0.9343
JHB 17	whole rock	19.52	0.8178
JHB 18	whole rock	23.32	0.8284
JHB 2	whole rock	22.23	0.8135
JHB 3	whole rock	22.22	0.8352
JHB 3	K-feldspar	43.05	0.8967
JHB 103	whole rock	4.259	0.7303
JHB 88	whole rock	5.04	0.7306
JHB 96	whole rock	19.04	0.8203
JHB 6925*	whole rock	21.01	0.8011
JHB 6926*	whole rock	38.70	0.8664
JHB 6927*	whole rock	30.69	0.8264
BH 18	whole rock	1.609	0.7169
BH 18	K-feldspar	3.472	0.7287

* Data from Bass (1968).

Table 4 K: Ar data, Belize

Sample	Analysed material	%K	Vol. rg^{40}Ar*	Age, Ma
JHB 43	biotite	5.71	5.473×10^{-5}	227 ± 7
JHB 46	biotite	7.68	7.231×10^{-5}	223 ± 7
JHB 53	biotite	7.31	6.852×10^{-5}	222 ± 7
BH 18	biotite	7.31	7.189×10^{-5}	232 ± 5

* Volume in cm^3/g at s.t.p.
Decay constants: λ $^{87}Rb = 1.47 \times 10^{-11}$ per year; λ_β $^{40}K = 4.72 \times 10^{-10}$ per year, λ_e $^{40}K = 0.584 \times 10^{-10}$ per year, $^{40}K/K = 0.0119$ At.%

Table 5 Sample localities

A*	JHB 25	Mountain Pine Ridge Granite, SW of Augustine, 16° 56′ N, 89° 02′ W
B*	JHB 6	Mountain Pine Ridge Granite, SSE of Augustine, 16° 57′ N, 88° 58′ W
K*	JHB 17	Mountain Pine Ridge Granite, ENE of Augustine, 17° 01′ N, 88° 52′ W
L*	JHB 18	Mountain Pine Ridge Granite, ENE of Augustine, 17° 00′ N, 88° 53′ W
M*	JHB 2	Mountain Pine Ridge Granite, E of Augustine, 16° 58′ N, 88° 56′ W
C*	JHB 3	Mountain Pine Ridge Granite, E of Augustine, 16° 58′ N, 88° 55′ W
D*	JHB 103	Mountain Pine Ridge Granite, NW of Baldy Beacon, 17° 01′ N, 88° 53′ W
H*	JHB 88	Bladen Porphyry, SE of Richardson's Peak, 16° 36′ N, 88° 44′ W
I*	JHB 96	Bladen Porphyry, SE of Richardson's Peak, 16° 35′ N, 88° 43′ W
J*	BH 18	Sapote Granite, WSW of Alabama, 16° 40′ N, 88° 38′ W
E*	JHB 43	Hummingbird Granite, SW of Middlesex, 16° 55′ N, 88° 32′ W
F*	JHB 46	Hummingbird Granite, SE of Middlesex, 16° 55′ N, 88° 27′ W
G*	JHB 53	Cockscomb Granite, SE of The Molar, 16°47′ N, 88° 33′ W

* Letters refer to localities on map (Figure 13).

has brought about changes in the Rb and Sr contents of the analysed material.

Bearing in mind the probably unsatisfactory nature of the analysed samples, the best-fit line in Figure 18A gives an age of 320±10 Ma with an intercept (initial ^{87}Sr:^{86}Sr ratio) of 0.7202± 0.003. Such an age (Late Carboniferous) is not unreasonable for this granite body.

Analyses of K-feldspars separated from two of the Mountain Pine Ridge samples plot significantly below the 320 Ma line (Figure 18A), which strongly suggests that movement of Rb and Sr between the different minerals of these granites has occurred. Such movement usually takes place in response to some later thermal event (or metamorphism), while recent weathering may have further accentuated these changes.

K: Ar data from the other granitic bodies provide evidence of a later, possibly Triassic, thermal event affecting the Maya Mountain block, while Clemons and Long (1971) show conclusively that late Cretaceous/early Tertiary plutonism occurred in the Central Cordillera of adjacent Guatemala.

Other granitic intrusives. A few determinations have been made on the Hummingbird and Cockscomb batholiths, and the Sapote Granite, a small stock to the south of the Cockscomb batholith; the petrographic and geochemical affinities between these rocks have been noted by Kesler and others (1974).

K: Ar determinations on biotites from these granitic rocks are given in Table 4. Within the limits of experimental error the apparent ages are identical, viz. 227±7 Ma (JHB 43, Hummingbird), 223±7 Ma (JHB 46, Hummingbird), 227±7 Ma (JHB 53, Cockscomb), and 237±5 Ma (BH 88, Sapote). Rb: Sr analyses on a sample of the Sapote Granite and on its K-feldspar give an age of 428±41 Ma with an initial ^{87}Sr:^{86}Sr ratio of 0.7068±0.0016. It is difficult to assess this age. The very limited palaeontological evidence suggests that the folded pre-Cretaceous rocks of the Maya Mountain Block are Carboniferous and Permian and there is no palaeontological support for an Ordovician–Silurian age for the rocks intruded by the Sapote Granite, as implied by the Rb: Sr data.

At this stage of investigation this result appears to be anomalous. Evidence has already been cited to show that the K-feldspars in

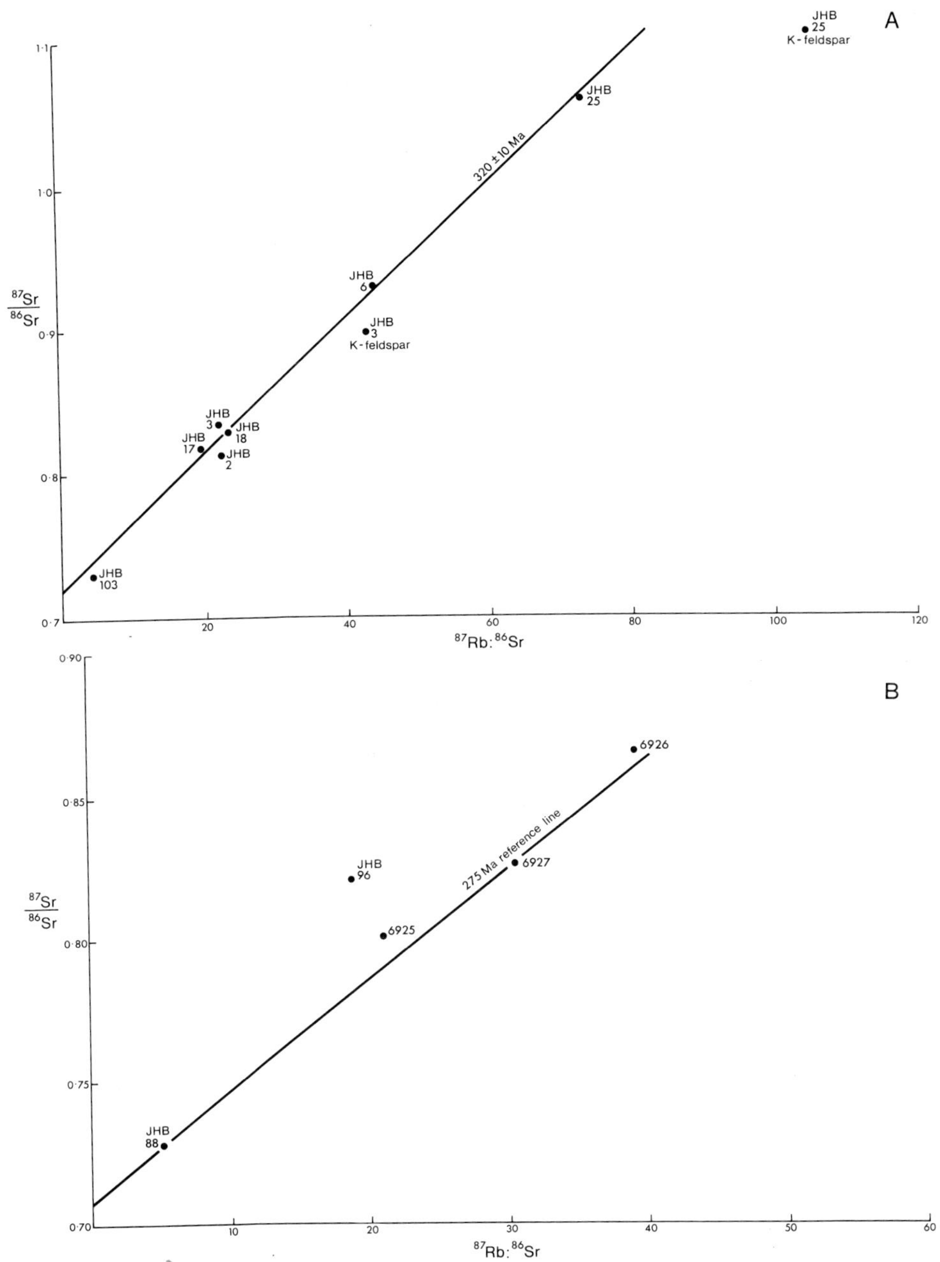

Figure 18

A Isochron diagram showing Rb: Sr data for whole rocks and K-feldspars from the Mountain Pine Ridge Granite.

B Isochron diagram showing Rb: Sr data for the whole rocks from the Bladen Volcanic Member. The 275-Ma reference line assumes an intercept on the ^{87}Sr:^{86}Sr axis of 0.71.

the Mountain Pine Range granites have suffered disturbance of their Rb:Sr systems and possibly the Rb:Sr relationships in the Sapote Granite sample have been similarly disturbed. The excellent concordance of the biotite K:Ar ages is taken as prima facie evidence of actual intrusion at about 225 Ma, that is during the Triassic.

The Bladen Volcanic Member. A late-Palaeozoic age can be assigned to these acid volcanics on the basis of their stratigraphic position. Utilising three Rb:Sr whole-rock analyses, Bass (1968) suggested an age of 280 Ma (his quoted age has been recalculated using the decay-constant for λ ^{87}Rb of 1.47×10^{-11} per year adopted in this report). Regression of the original data gives an 'age' of 242 Ma, but scatter of the data-points about the best-fit line is far greater than that which could be attributed to experimental error alone. Two additional samples have been analysed in the course of this investigation and all the data are plotted in Figure 18B; however, it is still clearly impossible to draw a straight line through these points to give a good fit; thus these volcanic rocks exhibit open-system behaviour with respect to Rb and Sr. Hence the radiometric data can offer no refinement of the stratigraphically-based estimate of the age of this unit.

Synopsis

Rb: Sr whole-rock analyses indicate an age for the Mountain Pine Ridge batholith of 320±10 Ma (late Carboniferous). Argon ages on biotites from the Hummingbird, Cockscomb and Sapote granites are identical within limits of experimental error at about 225 Ma, which is taken to indicate the age of intrusion. Attempts to date the Bladen Volcanic Member by Rb: Sr analyses were unsuccessful.

References

Bass, M. N. 1968. Age of the Bladen Porphyry, British Honduras. Pp. 79–80 *in* Variations in the isotopic abundances of strontium, calcium and argon, and related topics, by P. M. Hurley and others. *16th Annu. Prog. Rep. Dep. Geol. Geophys., Massachusetts Inst. Technol.* U.S. At. Energy Comm. Contract No. AT(30–1)–1381–16.

Bateson, J. H. 1972. New interpretation of geology of Maya Mountains, British Honduras. *Bull. Am. Assoc. Pet. Geol.*, Vol. 56, pp. 956–963.

— and Hall, I. H. S. 1971. Revised geologic nomenclature for pre-Cretaceous rocks of British Honduras. *Bull. Am. Assoc. Pet. Geol.*, Vol. 55, pp. 529–530.

Clemons, R. E., and Long, L. E. 1971. Petrologic and Rb: Sr isotopic study of the Chiquimula pluton, southeastern Guatemala. *Bull. Geol. Soc. Am.*, Vol. 82, pp. 2729–2740.

Hall, I. H. S., and Bateson, J. H. 1972. Late Palaeozoic lavas in Maya Mountains, British Honduras, and their possible regional significance. *Bull. Am. Assoc. Pet. Geol.*, Vol. 56, pp. 950–956.

Kesler, S. E., Kienle, C. F., and Bateson, J. H. 1974. Tectonic significance of intrusive rocks in the Maya Mountains, British Honduras. *Bull. Geol. Soc. Am.*, Vol. 85, pp. 549–552.

INDEX

Her Majesty's Stationery Office

Government Bookshops

49 High Holborn, London WC1V 6HB
13a Castle Street, Edinburgh EH2 3AR
41 The Hayes, Cardiff CF1 1JW
Brazennose Street, Manchester M60 8AS
Southey House, Wine Street, Bristol BS1 2BQ
258 Broad Street, Birmingham B1 2HE
80 Chichester Street, Belfast BT1 4JY

Government publications are also available through booksellers

Institute of Geological Sciences

Exhibition Road, London SW7 2DE

Murchison House, West Mains Road, Edinburgh EH9 3LA

The full range of Institute publications is displayed and sold at the Institute's Bookshop at the Geological Museum, Exhibition Road, London SW7 2DE

The Institute was formed by the incorporation of the Geological Survey of Great Britain and the Museum of Practical Geology with Overseas Geological Surveys and is a constituent body of the Natural Environment Research Council

Printed in England for Her Majesty's Stationery Office by Oxford University Press

Dd 496309 K10